AF733271

The Berlin Academy in the Reign of Frederick the Great

PHILOSOPHY AND SCIENCE

Oxford University Studies in the Enlightenment – formerly *Studies on Voltaire and the Eighteenth Century* (*SVEC*), is dedicated to eighteenth-century research.

General editor
Gregory S. Brown, University of Nevada, Las Vegas

Associate editors
Jenny Mander, University of Cambridge
Alexis Tadié, Sorbonne Université

Ex-officio
Nicholas Cronk, University of Oxford

Editorial board
Laura Auricchio, Fordham College
Katherine Brading, Duke University
Jean-Luc Chappey, Université de Paris Panthéon-Sorbonne
Andrew Curran, Wesleyan University
Elisabeth Décultot, Martin-Luther-Universität Halle-Wittenberg
Dan Edelstein, Stanford University
Nathalie Ferrand, École normale supérieure
Amy Freund, Southern Methodist University
Emily Friedman, Auburn University
Charlotte Guichard, Centre national de la recherche scientifique
Melissa Hyde, University of Florida
Catherine Jaffe, Texas State University
Andrew Jainchill, Queen's University, Canada
Andrew Kahn, University of Oxford
Lawrence Klein, University of Cambridge
Avi Lifschitz, University of Oxford
Antoine Lilti, École des hautes études en sciences sociales
Christophe Martin, Sorbonne Université
Pierre Musitelli, École normale supérieure
Christy Pichichero, George Mason University
Kate Quinsey, University of Windsor
Glenn Roe, Sorbonne Université
Ayana O. Smith, Indiana University
Céline Spector, Sorbonne Université
Karen Stolley, Emory University
Geoffrey Turnovsky, University of Washington
Stephane Van Damme, École normale supérieure

The Berlin Academy in the Reign of Frederick the Great
Philosophy and Science

Edited by

TINCA PRUNEA-BRETONNET

and

PETER R. ANSTEY

Published by Liverpool University Press on behalf of
© 2022 Voltaire Foundation, University of Oxford
ISBN 978 1 80207 025 5
Oxford University Studies in the Enlightenment 2022:11
ISSN 2634-8047 (Print)
ISSN 2634-8055 (Online)

Voltaire Foundation
99 Banbury Road
Oxford OX2 6JX, UK
www.voltaire.ox.ac.uk

A catalogue record for this book is available from the British Library

The correct style for citing this book is
Tinca Prunea-Bretonnet and Peter R. Anstey, eds, *The Berlin Academy in the reign of Frederick the Great: philosophy and science*
Oxford University Studies in the Enlightenment
(Liverpool, Liverpool University Press, 2022)

Cover illustration: Vignette from the title page of the first volume of the *Histoire de l'Académie royale des sciences et des belles lettres de Berlin* (Berlin, Haude, 1746). The National Library of Scotland, call number NE.797. Used with permission.

Printed and bound in Great Britain by TJ Books Limited, Padstow, Cornwall

Oxford University Studies in the Enlightenment

The Berlin Academy in the reign of Frederick the Great

PHILOSOPHY AND SCIENCE

This collection sheds new light on the nature, role and practice of philosophy and science in the renewed Berlin Academy from the mid-1740s to the 1770s, and in so doing provides a robust new instalment of materials for the broader task of constructing a historiography of philosophy at this important Enlightenment institution. The collection ranges from discussions of the roles of philosophy and natural philosophy in the formation of the reinvigorated Academy in the mid-1740s, to conceptions of the correct philosophical methodology to be deployed by the Academy. It provides the first ever study of the nature and arrangement of the new classes of the Academy, and a fresh appraisal of the Academy's methodological eclecticism. One recurring theme is the status of metaphysics: there are studies of special metaphysics, including the study of the soul; general metaphysics, that is, the study of being in general; and foundational metaphysical principles and concepts, such as Maupertuis's Principle of least action, Euler's concept of space and Lambert's notion of an experimental metaphysics. The collection also takes the study of the Academy in new directions through focused studies of important figures whose writings deserve to be better understood, such as Jean Bernard Merian, Louis de Beausobre, Jean Henri Samuel Formey and Johann Georg Sulzer.

Contents

Figures

Acknowledgements

This volume derives from a workshop on 'Speculative and experimental philosophy at the Berlin Academy' held at New Europe College and the Institute for Research in Humanities at the University of Bucharest in June 2016. While only four of the chapters included in the volume derive from this workshop – those of Peter Anstey, Paola Basso, Ansgar Lyssy and Tinca Prunea-Bretonnet – it was not long before the editors were able to procure an additional four chapters by scholars who were working simultaneously on cognate topics, namely, Angela Ferraro, Annelie Grosse, Christian Leduc and Alessandro Nannini. We would like to extend our thanks to New Europe College and the team at the Research Institute of the University of Bucharest for their support. The workshop was funded by UEFISCDI-CNCS (grant number PN-RU-TE-II 2014-4-2522), and Peter Anstey's attendance at the workshop was supported by his Australian Research Council Future Fellowship (FT120100282).

We would like to thank: the editors of *Philosophiques* for permission to reproduce the material published in 'L'économie de la nature: Maupertuis et Euler sur le principe de moindre action', *Philosophiques* 42:1 (2015), p.31-51, which appears in Ansgar Lyssy's chapter, 'Maupertuis, Euler and the Leibnizian metaphysics behind the Principle of least action'; the National Library of Scotland for permission to reproduce the vignette from the title page of the first volume of the *Histoire de l'Académie royale des sciences et des belles lettres de Berlin* (Berlin, Haude, 1746), call number NE.797; and the Niedersächsische Staats- und Universitätsbibliothek Göttingen for permission to reproduce figure IV from Johann H. Lambert's *Theorie der Zuverläßigkeit der Beobachtungen und Versuche, in Beyträge zum Gebrauche der Mathematik und deren Anwendung* (Berlin, Buchladen der Realschule, 1792), shelfmark 8 MATH I, 2949: 1.

A number of scholars and friends have provided invaluable feedback and advice on various aspects of this collection, including Karin de Boer, Fui Lee Luk, Sophie Roux, Eric Watkins and John Zammito. We are most grateful for their contributions.

Introduction

TINCA PRUNEA-BRETONNET
and PETER R. ANSTEY

The significance of the Berlin Academy in the reign of Frederick the Great (1712-1786) for our understanding of the Enlightenment has long been appreciated. It is fair to say, however, that, in anglophone scholarship at least, the nature of the Academy's contribution to philosophy, including its understanding of the discipline and its articulation with cognate fields such as natural philosophy, mathematics and the arts, is not well understood. In short, the historiography of philosophy at the renewed Berlin Academy is in a fluid state, and scholarly work on the subject remains rather piecemeal and disjointed.

To be sure, a series of recent biographical studies has gone some way to remedying this situation. Mary Terrall's study of Pierre-Louis Moreau de Maupertuis in *The Man who flattened the earth* (2002), which appeared a decade after David Beeson's *Maupertuis: an intellectual biography*, Ronald Calinger's *Leonhard Euler: mathematical genius in the Enlightenment* (2016) and more thematic studies such as Avi Lifschitz's *Language and the Enlightenment* (2012) have highlighted the importance of the Academy within the broader and enormously rich matrix of European discussion regarding issues of natural philosophy, natural theology and the origins and uses of language.[1] Again, the Academy appears in broad-ranged intellectual histories such as J. B. Shank's *The Newton wars* (2008) and Jonathan Israel's

1. Mary Terrall, *The Man who flattened the earth: Maupertuis and the sciences in the Enlightenment* (Chicago, IL, 2002); David Beeson, *Maupertuis: an intellectual biography*, *SVEC* 299 (1992); Ronald S. Calinger, *Leonhard Euler: mathematical genius in the Enlightenment* (Princeton, NJ, 2016); Avi Lifschitz, *Language and the Enlightenment: the Berlin debates of the eighteenth century* (Oxford, 2012).

Democratic Enlightenment (2011),[2] and there are articles on a variety of specific philosophical themes, such as scepticism at the Academy,[3] the Newtonianism of various of its leading members,[4] the monad controversy[5] and the status of metaphysics and aesthetics.[6] Some of this work has been undertaken using the somewhat anachronistic post-Kantian rationalism/empiricism distinction, and much of it focuses on leading figures such as Pierre-Louis Moreau de Maupertuis, Leonhard Euler and Jean Henri Samuel Formey and French philosophers who 'parachuted in', such as Julien Offray de La Mettrie. However, to date there are few if any studies on the nature and place of the experimental/speculative distinction in the Academy, on methodological diversity within the Academy, on natural theology or moral philosophy.

This collection aims to shed new light on the nature, role and practice of philosophy in the early renewed Academy from the mid-1740s to the 1770s with a view to providing a resource, a kind of robust new instalment, of materials for the broader task of constructing a historiography of philosophy at the Berlin Academy that is so sorely needed. In order to get our bearings for this task, this introduction provides not only a summary of the chapters that compose the

2. J. B. Shank, *The Newton wars and the beginning of the French Enlightenment* (Chicago, IL, 2008); Jonathan Israel, *Democratic Enlightenment: philosophy, revolution, and human rights 1750-1790* (Oxford, 2011), esp. p.188-95.
3. John C. Laursen, 'Tame skeptics at the Prussian Academy', *Libertinage et philosophie au XVIIe siècle* 12 (2010), p.219-28, and Nicolas Correard, 'Reasonable scepticism in the French Enlightenment: some connections between Jean-Baptiste Boyer d'Argens, Louis de Beausobre, and Voltaire', in *Scepticism in the eighteenth century: Enlightenment, Lumières, Aufklärung*, ed. Sébastien Charles and Plínio J. Smith (Dordrecht, 2013), p.173-88.
4. Thomas Ahnert, 'Newtonianism in early Enlightenment Germany, c. 1720 to 1750: metaphysics and the critique of dogmatic philosophy', *Studies in history and philosophy of science* 35:3 (2004), p.471-91.
5. Thomas Broman, 'Metaphysics for an enlightened public: the controversy over monads in Germany 1746-1748', *Isis* 103:1 (2012), p.1-23.
6. Ahnert, 'Newtonianism'; Christian Leduc, 'La métaphysique de la nature à l'Académie de Berlin', *Philosophiques* 42:1 (2015), p.11-30; and Stefanie Buchenau, '"L'œil ne se voit pas voir": Sulzer sur la contemplation et le sentiment de soi', *Philosophiques* 42:1 (2015), p.73-88. See also Udo Thiel on Merian: 'Between Wolff and Kant: Merian's theory of apperception', *Journal of the history of philosophy* 34 (1996), p.213-32, and *The Early modern subject: self-consciousness and personal identity from Descartes to Hume* (Oxford, 2014), part V.

volume, but also a summary history of the Berlin Academy from its inception until the death of Frederick the Great.[7]

After years of effort and a number of false starts, in 1700-1701 the German polymath Gottfried Wilhelm Leibniz founded the Society of the Sciences (Sozietät der Wissenschaften) in Berlin with the explicit aim of following the examples of the Royal Society of London (founded in 1660) and both the Académie française (founded in 1635) and the Parisian Academy of Sciences (founded in 1666). It was not the first German-speaking academy; the Academia Naturae Curiosorum or Leopoldina had been founded in 1652 in Schweinfurt, after the model of Italian academies, and had become an important institution, especially after its recognition as an imperial academy in 1687. Nevertheless, the Society of the Sciences was the first serious attempt to establish a society rivalling the success and importance of those in London and Paris, with the additional aim of promoting the arts.

The founding of the Berlin Academy came only a few years after the opening of the University of Halle and the creation of the Academy of Arts in Berlin (1696), during the reign of Frederick I and with the significant support of Sophie Charlotte. Its composition reflected the important role played in the intellectual life of the period by the Huguenots established in Berlin in the 1680s where they had opened the Collège français (1689). The Berlin Academy established a strong and long-lasting relationship with this community, hiring several of the teachers active at the Collège as its members, and encouraging prominent scholars, physicians and theologians of French origin. The Huguenot community would be significant for the later development of the institution too, as Formey, for instance, occupied the position of permanent secretary of the Academy for almost half a century (1748-1797).[8]

Despite several well-known members, such as Daniel Ernst Jablonski, Friedrich Hoffmann, Johann Leonhard Frisch, August

7. Other summary histories of the Berlin Academy can be found in Christian Leduc and Daniel Dumouchel, 'Introduction', *Philosophiques* 42:1 (2015), p.7-10; Terrall, *The Man who flattened the earth*, ch.8; and Calinger, *Leonhard Euler*, ch.6. For the Academy's official history, see *Histoire de l'Académie royale des sciences et belles lettres, depuis son origine jusqu'à présent, avec les pieces originales* (Berlin, Haude and Spener, 1750), p.1-81.
8. For the Huguenots in Berlin, see *The Berlin Refuge, 1680-1780: learning and science in European context*, ed. Sandra Pott, Martin Mulsow and Lutz Danneberg (Leiden, 2003).

Hermann Francke, Johann and Jakob Bernoulli, Nicolaas Hartsoeker, Christian Wolff and Kaspar Neumann, and despite a certain openness to the European intellectual scene, many of the Academy's members had a mostly regional impact during its first three decades.[9] The first decade was characterised by organisational challenges, Leibniz being often absent from Berlin. Still, the Academy managed to publish the first volume of its proceedings in 1710 under the title *Miscellanea Berolinensia*, though only seven issues were published, the last in 1743.[10]

Previous histories of the Berlin Academy, such as the monumental work of Adolf von Harnack, have emphasised Leibniz's crucial role in the organisation of the institution and his personal views on its functioning.[11] Recent scholarship, however, has been more contextual, highlighting the significance of the Parisian and London models which inspired him. Leibniz's wish to erect on German soil a comparable institution able to collaborate, if not to compete, with these two prestigious academies was revived and reinforced by its later reorganisation under Frederick the Great.

After Frederick William I ascended to the throne in 1713, the Academy experienced a period of relative neglect, the king being more focused on politics and warfare and less supportive of culture and science in general. He did, however, show interest in medicine and botany, for instance, as well as in sciences that could bring immediate practical results and demonstrate their utility. Despite the increasing number of members, the Academy suffered from a lack of financial support and recognition, especially after Leibniz's death in 1716, and did not manage to reach the renown the latter had hoped this institution would achieve. This situation, however, changed drastically after the ascension of Frederick II to the throne in June 1740.

In the very month of his accession, the new king sent invitations to famous scholars he hoped to bring to Berlin and with whom he envisaged starting the reorganisation of the Academy. Some of them, such as Willem 's Gravesande and Jacques de Vaucanson, refused, but others, including Leonhard Euler and Francesco Algarotti, did respond positively. Voltaire, whom the king had tried to coax to Prussia in previous years, did not permanently move to Berlin, despite

9. See Conrad Grau, *Die Preussische Akademie der Wissenschaften zu Berlin* (Heidelberg, 1993), p.77.
10. *Histoire de l'Académie royale depuis son origine*, p.40.
11. See Grau, *Die Preussische Akademie*, p.55-57.

several visits, but actively supported this reorganisation. Frederick the Great wished to give the Academy the renown and breadth worthy of an important kingdom, and wanted this institution to express his cultural, scientific and political views and ambitions. In the early years, both Formey and Euler made significant contributions to Frederick's plans for his new Academy. His most strategic move, however, was to lure Maupertuis, an eminent natural philosopher, a member of several academies and well known throughout Europe for having proved Newton's theory regarding the shape of the earth during a highly publicised expedition to Lapland.[12] Frederick saw in him a character able to succeed in the ambitious task of transforming his academy into one of the most important institutions in Europe. They were to have a harmonious collaboration, based on trust, respect and friendship until Maupertuis's death in 1759.

After a few years of hesitation and a series of structural changes,[13] the renewal of the renamed Royal Academy of Sciences and Belles-Lettres was completed in May 1746 with Maupertuis as its all-powerful president. Four classes were established (mathematics, experimental philosophy, speculative philosophy and belles-lettres), all sessions were common and an annual volume of proceedings was to profile the works of the members. The activity of the Academy was no longer directly connected to practical outcomes: the progress of science, philosophy, and literature and history, even of reason itself, became its goals,[14] in harmony with Frederick's vision.

Of particular importance in the revivified Academy was the inclusion of the class of speculative philosophy, for which there was no parallel elsewhere in Europe. It was to include logic, metaphysics and morals within its ambit, and it is among the outputs of this class that much of the philosophical interest in the Academy lies. The inclusion of a class of belles-lettres also set the Berlin Academy apart from those in Paris and London, as Maupertuis stressed in his 'On the duties of the academician': 'this company embraces a field way vaster than most other academies.'[15] The breadth of the Academy's activities

12. See Terrall, *The Man who flattened the earth*.
13. See Peter Anstey's chapter in this volume.
14. Wilhelm Dilthey, 'Friedrich der Grosse und die deutsche Aufklärung', in *Studien zur Geschichte des deutschen Geistes*, *Gesammelte Schriften*, vol.3 (Leipzig and Berlin, 1927), p.81-157 (144).
15. Pierre-Louis Moreau de Maupertuis, 'Des devoirs de l'académicien', in *Histoire de l'Académie royale depuis son origine*, p.115-25 (117), reprinted in *Œuvres de*

reflects the unified nature of Frederick's vision, as much as the pragmatics of grouping all intellectual pursuits under the one organisation in a metropolis as small as mid-eighteenth-century Berlin, whose population has been estimated to be around 90,000 in 1750.[16]

The king also succeeded in co-opting some of Europe's most important scholars as corresponding members of the Academy, such as Etienne Bonnot de Condillac. However, significant German thinkers, such as Christian Wolff and Albrecht von Haller, declined his invitation to join as full members. Full members had to reside in Berlin, and many of them were French-speaking intellectuals. Swiss savants, such as Jean Bernard Merian, Nicolas de Béguelin and Johann Georg Sulzer, came to play decisive roles in its functioning.[17] This situation reflected not only Frederick's preference for the French language, but, above all, his willingness to encourage an intellectual model where abstract and scientific knowledge was accompanied by literary style, formal beauty and intelligibility. He regarded Voltaire, Maupertuis and Diderot as exemplars to be recommended to German thinkers.

Needless to say, Frederick wanted his Academy to be famous throughout Europe, and its activities to be followed by the Republic of Letters. This is one of the reasons why he decided that French should be its official language, and indeed until 1793, when a parallel series was published in German (entitled *Sammlung der deutschen Abhandlungen*), the Berlin Academy would publish its *mémoires* only in French. The king also believed that French language and literature would help refine the German approach to style and writing and perfect the German language.

Earlier in the century, Frederick I had encouraged the Academy to work on a dictionary of German terms, in order to contribute to the development of the German language,[18] as well as to the history of the German states. Frederick the Great believed deeply that German

Maupertuis, 4 vols (Lyon, Bruyset, 1768), vol.3, p.283-302; see the appendix to this volume. All translations are our own unless otherwise stated.

16. See Erik Hornung, 'Diasporas, diversity, and economic activity: evidence from 18th-century Berlin', Centre for Competitive Advantage in the Global Economy (University of Warwick), Working Paper Series, no.390 (2018), https://warwick.ac.uk/fac/soc/economics/research/centres/cage/manage/publications/390-2018_hornung.pdf (last accessed 10 May 2022), p.6.

17. See *Schweizer im Berlin des 18. Jahrhunderts*, ed. Martin Fontius and Helmut Holzhey (Berlin, 1996).

18. Adolf von Harnack, *Geschichte der Königlich Preussischen Akademie der Wissenschaften zu Berlin*, 3 vols (Berlin, 1900), vol.1.1, p.129.

needed to be improved,[19] and that this could be achieved by following the example of the French language and first of all by supporting translations from ancient and modern authors. Knowledge had to be articulated in such a way as to be understood by all citizens.[20] To be sure, he regarded the Berlin Academy as a 'university' for the noble class and for the officials of the Prussian state,[21] yet he also advocated the ideal that the people as such should be enlightened. If the taste of the public is refined through education and high-quality reading material (such as good translations of ancient works), he argued, new authors would also write in an elegant yet comprehensible language and the German philosophers and *génies* would thereby find their much-needed 'Prometheus' able to animate and enliven their ideas.[22]

Frederick himself, who deeply admired both Marcus Aurelius and Louis XIV, regarded himself as endorsing the role of a 'Medici', of the enlightened ruler supporting the progress of philosophy, the sciences and the arts by all the means at his disposal,[23] and primarily through his Academy. Moreover, he wished to highlight not only the grandeur of the Prussian state, but also the freedom of thought he was committed to encouraging, happily welcoming to Berlin thinkers who were outlawed in their native countries, such as the French exile La Mettrie. Frederick also actively participated in the intellectual activity of the Academy, by writing and publishing discourses and essays, though without being present at its sessions. He remained directly involved in the choice of new members, especially after Maupertuis's death in 1759 when Frederick became the official president of the Academy. Euler unofficially took over the late Maupertuis's duties from 1759; however, after Euler's departure for Russia in 1766, Jean D'Alembert's role in advising the king in academic issues became increasingly important. D'Alembert nevertheless refused to take up the presidency of the Academy in spite of Frederick's invitation.

19. Frederick II, king of Prussia, 'De la littérature allemande', in *Œuvres philosophiques* (Paris, 1985), p.423-55 (434).
20. Frederick II, 'De la littérature', p.454: 'Scholars should remember sometimes that sciences are food for the soul: memory reacts like the stomach; but they cause indigestion if judgement does not digest them. If our knowledge is a treasure, we should not bury it, but make it fertile by disseminating it in a language understood by our fellow citizens.'
21. See Dilthey, 'Friedrich der Grosse', p.117, who calls it 'a modern university for the ruling class'.
22. Frederick II, 'De la littérature', p.440-51.
23. Frederick II, 'De la littérature', p.455.

The main means by which the Academy influenced European intellectual life were the prize essay questions published annually by one of its classes, a practice common amongst European academies. Three languages were accepted for the essays: French, German and Latin. Their impact on the Republic of Letters was enormous, and scholars eagerly awaited the chosen questions – especially those of the speculative philosophy class[24] – as an indication and expression of the most pressing and significant matters of the day. Indeed, the Academy was very strategic in the manner in which it promoted contemporary debates, encouraged scholars to look for solutions, and tried to guide philosophical reflection towards the problems it regarded as central or fertile but neglected. Numerous intellectuals from all over Europe responded with essays to these prize questions, and the Academy published the essays that won a prize or an *accessit*. Needless to say, the *Preisfragen* elicited other writings aiming to contribute to the general discussion, beyond the academic contests.

Several of the philosophical and literary prize questions were particularly hotly debated, such as those regarding Leibnizian (and Wolffian) philosophy in 1747, 1751 and 1755 (on monads,[25] free will and the best possible world), method in philosophy in 1763, the *éloge* of Leibniz in 1768 and the question of inclinations and fundamental powers of the soul in 1767-1768 and 1776. Just as significant were the prize questions regarding the origin of languages in 1771, the problem of taste in 1775 and the role of the French language in 1784.[26] In 1780 the Academy accepted a question strongly supported by Frederick II (and by D'Alembert, its initiator): 'Is it useful for a nation to be deceived, whether by leading it into new errors or by maintaining old ones?', which stimulated responses right across Europe.[27]

Many famous thinkers won the prize or the *accessit* during Frederick's reign, such as D'Alembert, Mendelssohn, Herder, Kant, Kästner, Eberhardt, Garve and Michaelis. Sometimes, the Academy chose less significant essays, such as those by J. H. G. Justi and

24. Harnack, *Geschichte*, vol.1.1, p.396.
25. On the monad controversy, see Broman, 'Metaphysics'.
26. For details, see Cornelia Buschmann, 'Die philosophischen Preisfragen und Preisschriften der Berliner Akademie der Wissenschaften im 18. Jahrhundert', in *Aufklärung in Berlin*, ed. Wolfgang Förster (Berlin, 1989), p.165-227. For a list of the prize questions and the winners, see Harnack, *Geschichte*, vol.2, p.305.
27. See Hans Aarsleff, 'The Berlin Academy under Frederick the Great', *History of the human sciences* 2:2 (1989), p.193-206 (194-95).

A. F. Reinhard, and in these cases its judgement was subjected to critical scrutiny by contemporary thinkers. Interestingly, many of the best contested and most publicised prize questions were from the class of speculative philosophy, and it is the activity and contribution of this class to the Enlightenment that is the primary focus of this volume.

The chapters in this collection range from discussions of the roles of philosophy and natural philosophy in the formation of the reinvigorated Academy in the mid-1740s to conceptions of the correct philosophical methodology to be deployed by the Academy. To some extent, the volume builds upon the important special issue of *Philosophiques* of 2015 that explored particular philosophical themes in detail, including Johann Georg Sulzer's aesthetics,[28] Maupertuis and Euler on the Principle of least action,[29] and the methodological debates between the Wolffians and the Newtonians in the Academy.[30] Thus, one recurring theme in the current collection is the status of metaphysics. There are studies of both particular metaphysics, that is, the study of the soul, and general metaphysics, that is, the study of being and foundational metaphysical principles. One such principle that comes in for special treatment is Maupertuis's Principle of least action, as does Euler's concept of space. Yet this collection also takes the study of the Academy in new directions.

In his chapter on 'The four classes of the Berlin Academy', Peter Anstey argues that the new names given to the Academy's four classes in 1746 reflect a decidedly British influence. He argues that the new names 'class of experimental philosophy' and 'class of speculative philosophy' derive from the long-held distinction between experimental and speculative philosophy that was first deployed in the early Royal Society of London some eighty years earlier in the 1660s. What is distinctive in Berlin, however, according to Anstey, is the absence of any antagonism between the two classes within the Academy. Since the emergence of experimental philosophy in the latter half

28. Daniel Dumouchel, 'Entre métaphysique et esthétique: sentiment et beauté dans les mémoires de Johann Georg Sulzer', *Philosophiques* 42:1 (2015), p.51-72, and Buchenau, 'L'œil ne se voit pas voir'.

29. Ansgar Lyssy, 'L'économie de la nature: Maupertuis et Euler sur le principe de moindre action', *Philosophiques* 42:1 (2015), p.31-51.

30. Tinca Prunea-Bretonnet, 'La méthode philosophique en question: l'Académie de Berlin et le concours pour l'année 1763', *Philosophiques* 42:1 (2015), p.107-30, and François Duchesneau, 'Nicolas de Béguelin et les fondements d'une philosophie de la nature', *Philosophiques* 42:1 (2015), p.89-105.

of the seventeenth century, experimental philosophy had been set against speculative philosophy and, in particular, against the use of hypotheses and systems. By contrast, the situation in Berlin in the mid-1740s was such that not only was there an absence of any tension between these classes, but metaphysics within the class of speculative philosophy was actively promoted. Anstey argues that the high status of metaphysics within the revivified Academy is best understood by tracing the origins of the anti-system debate in Britain and Europe.

Annelie Grosse, in effect, picks up the theme of the status of metaphysics in the class of speculative philosophy and explores the tensions between its proponents. She argues that Formey, the secretary of the Academy, saw metaphysics as being grounded on *a priori* metaphysical principles such as the Principle of sufficient reason, whereas the president, Maupertuis, gave a central role to his Principle of least action. When it came to arguing for the existence of God, for example, Formey began with the Principle of sufficient reason. By contrast, Maupertuis developed a two-pronged proof, arguing that the fact that the laws of nature can be derived from the Principle of least action and from observation is compelling evidence for the existence of a divine being. Moving from Formey and Maupertuis to N. Béguelin and J. B. Merian, Grosse examines how these two philosophers continued to argue for the high status of metaphysics in the late 1750s and the years after the death of Maupertuis. She shows how Merian was keen to dissociate metaphysics from mathematics and plumped for a broader, more eclectic approach to the subject. In this he was consciously positioning himself on the question of the status of metaphysics as a response to the divergent views expressed in the Academy's prize essay contest, announced in 1761 and awarded in 1763, on the certainty of metaphysical truths. Formey, for his part, was never able to revise his conception of metaphysics as a demonstrative science based on first principles, in spite of the continued and vigorous debates on the subject in the Academy over two decades.

Tinca Prunea-Bretonnet's chapter on eclectic philosophy in the Academy uses Jean Bernard Merian's *Discourse on metaphysics* of 1765 as a key text for establishing the claims that a homogeneous and consistent academic philosophy arose in the Academy in the 1760s, and that it took the form of an eclectic methodology. Prunea-Bretonnet argues that this shared theoretical approach to philosophy replaced the previous opposition between the two main groups, one Wolffian and the other anti-Wolffian and Newtonian. This approach to philosophy is also evident in the methodological essays of the

Academy's secretary Formey from the period, and is even adumbrated in earlier writings by Maupertuis. It effectively ran in parallel with the broader concern with moral philosophy specific to other contemporary currents, and is characterised by a focus on a set of eclectic methodological principles, founded on the theses of the primacy and specificity of metaphysics. Prunea-Bretonnet finds the origins of this eclecticism in the earlier writings of Christian Thomasius, thus establishing a significant Thomasian legacy in the intellectual life of the Academy. Unlike Thomasius, however, who is critical of metaphysics, the academicians emphasise the crucial role of this science.

Angela Ferraro's chapter on the philosophy of Louis de Beausobre sets out his Wolffian methodology, which seeks a balance, a golden mean, between experience and reasoning. Ferraro shows how Beausobre was critical of both Emilie Du Châtelet's approach to experience and Condillac's attack on abstract principles in his *Treatise on systems* (1749). The thrust of Beausobre's position is to be sceptical about observation and anatomical and microscopical experiments. Indeed, he found Condillac's more mature *Treatise on sensations* more conducive with its emphasis on the need to build experience over time. He was also critical of fellow academician Johann Friedrich Meckel's anatomical approach to the study of madness, claiming that the locus of study should be the soul, which is to be approached via introspection. Yet how is the balance between reasoning, with its appeal to principles such as the Principle of sufficient reason, and experience, given the fallibility of experiments and observation, to be established in disciplines such as natural philosophy and psychology? On this point Ferraro helpfully contrasts Beausobre with J. B. Merian, and in so doing provides a nuanced interpretation of the former.

Ansgar Lyssy's chapter is concerned with the epistemic status of Maupertuis's Principle of least action and the manner in which it intersects with the two rival approaches to the metaphysics of forces in Newtonianism and Leibnizianism. This enquiry, in turn, provides a case study for a broader assessment of the general orientation of the Berlin Academy vis-à-vis these two conflicting methodologies. Lyssy's analysis shows that both Maupertuis and Euler grounded the Principle of least action on certain metaphysical doctrines deriving from Leibniz, including the economy of nature, an acceptance of final causes and their equivalence with efficient causes, and the conception of force. While Maupertuis attempted to ground the Principle on experience, commencing with optics, in contrast to Euler – whose approach was to follow a kind of neo-Euclidean method – both *philosophes* sided with a more Leibnizian, metaphysical approach in contrast to Maupertuis's Newtonian critic Patrick d'Arcy. Thus, Maupertuis's and Euler's approaches appear to be consistent with the sort of relation between natural philosophy and metaphysics that we find in the Academy in general.

Christian Leduc's chapter also concerns Euler, though his focus is on Euler's arguments for the origins of our idea of absolute space. Interestingly, Euler rejects the view that our knowledge of the nature of space is acquired through the senses, and claims instead, in his *Réflexions*, that it comes via reflection. Just what he means by 'reflection' is the central focus of the chapter. Leduc argues that, according to Euler, our idea of space is acquired neither via Lockean reflection nor via the form of reflection advanced by Condillac that amounts to a process of mental abstraction. Instead, it is acquired through an inferential process of positing theoretical oppositions to sensible properties such as impenetrability and motion. It is through a process of separation that we acquire the notions of immobility, penetrability and infinity. Surprisingly, given Euler's criticisms of other aspects of Leibniz's thought, Leduc finds some interesting parallels between this conception of reflection and that of Leibniz. This, according to Leduc, should hardly surprise us, given that there was a strong Leibnizian element within the milieu of the Academy.

The status of metaphysics is also a central theme in Paola Basso's chapter on the methodology of the triumvirate Maupertuis, Euler and Lambert. Basso explores what she calls the hybridisation of natural philosophy, metaphysics and mathematics in the thought of these leading figures of the Academy. Beginning with Lambert's speech before the Academy in 1765, she sets out his understanding of the manner in which experimental natural philosophy closely articulates with the classes of mathematics and speculative philosophy. Lambert's vision finds parallels in Euler and, interestingly, as Basso points out, Georg Johann von Holland used a striking term to describe this approach, namely, 'experimental metaphysics'. Perhaps the best exemplar of experimental metaphysics in the Academy is the teleological treatment of the Principle of least action by both Maupertuis and Euler.

The Principle of least action as discussed by Maupertuis and Euler pertains to general metaphysics, but some within the Academy were also concerned with particular metaphysics, which was focused on the human soul. In his chapter on Sulzer's writings on the soul,

Alessandro Nannini first sets out Sulzer's account of how the soul is related to the body. According to Sulzer, the soul, whose seat is the brain, is an active force whose powers can only be activated when situated in a human body; otherwise it is completely inactive. The sense organs mediate the multifarious sources of information about the external world via the nerves in such a way that the soul is able to exercise its faculty of knowing. Yet humans have not always enjoyed the rich epistemic and aesthetic states that they are now in. Sulzer theorises as to the process through which we, beginning as savages, travelled in order to reach our present condition. Moving from passivity to the ability to discriminate between things in our environment, we eventually developed the mechanical arts and then began to congregate in societies. Then came the sciences and fine arts, and Nannini devotes the remainder of his chapter to Sulzer's fascinating account of the development of the *beaux-arts*, drawing on the Orpheus myth and giving a prominent place to music. On Nannini's reading, Sulzer's process of humanisation is a delicate balancing of *aisthesis* and *anaisthesis*.

The collection concludes with an appendix, which contains a fresh English translation of Maupertuis's 'On the duties of the academician', first published in the *Histoire de l'Académie royale des sciences et belles lettres, depuis son origine jusqu'à présent, avec les pieces originales* in 1750. This short speech encapsulates Maupertuis's vision for the Academy and its members, and provides an insight into his understanding of the uniqueness of the organisation, the importance of its various classes, as well as its preference for French over German. Important variants in later editions of the speech are recorded in the notes. To our knowledge it is the first such translation since the loose rendering that appeared in *The Gentleman's magazine* in 1751.[31]

Having summarised the contents of the chapters that follow, it remains now to draw the reader's attention to two important points, one bibliographic and the other a point of translation. The published proceedings of the Berlin Academy are an extraordinary repository of Enlightenment thought. They contain not only detailed histories of the institution, but scores of papers of merit, some being outstanding contributions to knowledge in their own right. Ranging across the four classes of the Academy, the proceedings' contents are an unparalleled

31. See *The Gentleman's magazine and historical chronicle* 21 (May 1751), p.218-22.

resource for Prussian intellectual life from the mid-1740s to the end of the century. Nevertheless, there are complications in citing from them. First, from 1746 until 1769 their official title, *Histoire de l'Académie royale des sciences et belles lettres* (*Histoire de l'Académie royale des sciences et des belles lettres de Berlin* in 1745), is almost identical to that of the main publication of the Parisian Academy in the same period, namely, *Histoire de l'Académie royale des sciences.* This can lead to problems of disambiguation. Second, the title of the proceedings changed twice: first, in 1770 it became *Nouveaux mémoires de l'Académie royale des sciences et belles-lettres*; and it changed again in 1786 to *Mémoires de l'Académie royale des sciences et belles-lettres.* Third, their contents are divided into 'History' and 'Memoirs' (i.e. published articles), and this is normally reflected by separate pagination for the memoirs in the same volume. Fourth, normally the date of publication postdates the date of oral presentation of a particular article by two years or more, so that a paper read, say, in 1748 did not appear in print until 1750. In short, citing from the Academy's proceedings is as difficult as it is necessary.

Finally, in the preparation of this current volume on the Berlin Academy, the question arose as to the correct translation of the French noun *physique*, and it is worth highlighting for the reader the issues involved and stating how this has been handled in the pages that follow. The standard French term for the science of nature in the period was *physique*, which derives ultimately from the ancient Greek word for nature, namely, φύσις. The standard English term for the science of nature in the seventeenth and eighteenth centuries was 'natural philosophy', though there was a handful of variants and alternatives. These variants include words that are also ultimately derived from the Greek φύσις, such as 'physiology', 'physick', 'physicks' and 'physics'.

It might seem natural then to translate the French *physique* as 'physics', and indeed in some contexts a case can be made that this is the most appropriate translation. But there are complications. First, in early modern English 'physick' is the most common term for therapeutic medicine, and this most certainly is not coextensive with 'natural philosophy'. Second, in contemporary English the term 'physics' has a far narrower semantic domain than the term 'natural philosophy' or *physique.* Thus, if an eighteenth-century French writer is referring to Newton's writings in *physique*, they are often referring to Newton's writings in natural philosophy, including his *Opticks* and even his 'chymical' writings. However, to a modern reader the English expression 'Newton's writings in physics' would normally be taken to refer more narrowly to the celestial dynamics as set out in his *Principia*, including his laws of motion and gravitational attraction. Given these considerations, it seems most natural, unless contextual reasons override it, to translate ***physique*** as 'natural philosophy'.

Of course, we do not presume to have presented a thorough lexically based evidence set for this recommendation, and this is not the appropriate place for such evidence to be adduced. Nevertheless, we conclude with two well-known passages that serve to drive home the point: one from the final chapter of Pierre Coste's French translation of Locke's *Essay concerning human understanding* of 1700, and a second from the article 'Physique' in the *Encyclopédie*.

In the chapter on the division of the sciences in Pierre Coste's translation of Locke's *Essay*, we find the following translation: 'La première [espèce de science] est la connaissance des choses comme elles sont dans leur propre existence, dans leurs constitutions, propriétés et opérations [...] la matière et le corps, mais aussi les esprits [...] C'est ce que j'appelle ***physique*** ou ***philosophie naturelle***, en prenant ce mot dans un sens un peu plus étendu qu'on ne fait ordinairement.'[32] Note that Coste speaks of '***physique*** ou ***philosophie naturelle***', intimating that they are synonyms. Interestingly, Locke's original has 'φυσική, *or natural Philosophy*' (i.e. using the Greek term which Coste translates as ***physique***). Locke's self-consciously extended sense of 'natural philosophy' is not restricted to material things but includes spirits and God.[33]

Turning to D'Alembert's article on 'Physique' in the twelfth volume of the *Encyclopédie*, published in 1765, we find the following: 'PHYSIQUE, [...] cette science que l'on appelle aussi quelquefois *Philosophie naturelle*, est la science des propriétés des corps naturels, de leurs phénomenes & de leurs effets, comme de leurs différentes affections, mouvemens, *&c.*'[34] Once more we find the identification of ***physique*** with ***philosophie naturelle***, that is, natural philosophy, and the contents of this science are the same as we find in Locke, though without extending to spirits. It is these kinds of cases that lead us to

32. John Locke, *Essai philosophique concernant l'entendement humain*, translated by Pierre Coste (1700), ed. Philippe Hamou (Paris, 2009), IV.xxi.2, p.1037.
33. John Locke, *An Essay concerning human understanding* (1690), ed. Peter H. Nidditch (Oxford, 1975), p.720.
34. *Encyclopédie, ou Dictionnaire raisonné des sciences, des arts et des métiers*, ed. Denis Diderot and Jean D'Alembert, University of Chicago: ARTFL Encyclopédie Project (autumn 2017 edn), ed. Robert Morrissey and Glenn Roe, http://encyclopedie.uchicago.edu/ (last accessed 13 April 2022), vol.12, p.539.

claim 'natural philosophy' really is the most natural translation of *physique*.

We conclude with an example that highlights the importance of this point. In the first part of the *Discours préliminaire* of 1751, when referring to the utility of many discoveries about nature, D'Alembert claims: 'Voilà l'origine & la cause des progrès de cette vaste Science, appellée en général Physique ou Etude de la Nature, qui comprend tant de parties différentes.'[35] Richard Schwab, in his popular edition of the *Preliminary discourse*, translated this as follows: 'Such was the origin and the cause of progress of the vast science generally called Physics, or the Study of Nature, which includes so many different parts.'[36]

We contend, however, that 'physics' is not an appropriate translation for *physique* for a modern readership here. Agriculture and medicine were the first of those 'parts', according to D'Alembert – hardly the stuff of Newtonian 'physics' in the modern sense. Thus, this is a clear instance in which *physique* is best translated as 'natural philosophy', rather than 'physics' as preferred by Schwab. We could easily multiply instances, but enough has been said, in our view, to establish a case for translating *physique* as 'natural philosophy', though we should point out that our contributors in the following chapters have made their own decisions with regard to how the French term should be translated.[37]

35. *Encyclopédie*, vol.1, p.4.
36. Jean D'Alembert, *Preliminary discourse to the Encyclopedia of Diderot*, translated by Richard N. Schwab (Chicago, IL, 1995), p.16.
37. For a parallel situation in eighteenth-century German, see the entry 'Philosoph' in *Lexikon der Aufklärung: Deutschland und Europa*, ed. Werner Schneiders (Munich, 2001), p.303, where Helmut Holzhey identifies 'Physiker' with 'natural philosopher'.

The four classes of the Berlin Academy

PETER R. ANSTEY

It is well known that the Berlin Academy comprised four classes. In fact, the organisational structure, activities and outputs of the Academy in the period of its revival during the reign of Frederick the Great can only be understood in the light of these four classes. They determined the structure both of the membership and of the Academy's main publication during this period, the *Histoire de l'Académie royale des sciences et belles lettres* (hereafter *Histoire*). It might even be argued that the classes of the Academy in Berlin played a more important role than in any other European or British scientific society in the mid-eighteenth century. Nevertheless, to my knowledge, there has been little sustained study of the classes themselves, in terms of either the philosophical background that gave rise to them or their members and outputs.[1] In this chapter, I aim to make a modest contribution to the former issue by arguing that there was a decidedly British strand of influence in the formation of the classes themselves.

I will argue that the distinction between the class of experimental philosophy and the class of speculative philosophy is a direct

1. Symptomatic of the relative neglect of the Academy's classes is the way in which Mary Terrall, in her otherwise very good treatment of Maupertuis and the renewal of the Academy, conflates the classes of 1744 with those of 1746. See Mary Terrall, *The Man who flattened the earth: Maupertuis and the sciences in the Enlightenment* (Chicago, IL, 2002), p.238-39. It should be pointed out that the summary history of the Academy in the *Histoire* of 1748 elides the classes of 1744, moving from those of 1710 to those of 1746; see *Histoire de l'Académie royale des sciences et belles lettres* (1746; Berlin, Haude and Spener, 1748), p.1-2. For recent work on the class of speculative philosophy, see Christian Leduc, 'Speculative philosophy at the Berlin Academy', in *History of universities*, vol.33.2, ed. Mordechai Feingold and Andrea Sangiacomo (Oxford, 2020), p.175-200.

descendant of British experimental philosophy that emerged in England eight decades earlier in the 1660s. Further, I will argue that the status of metaphysics in the Enlightenment in general, and within the class of speculative philosophy in particular, has been misunderstood, at least in anglophone historiography of the Enlightenment. This is because it has been systematically conflated with the critique of systems. Of course, it was not simply the British influence that led to the establishment of the class of speculative philosophy in the renewed Berlin Academy of 1746; the influence of Leibnizian and Wolffian metaphysics was also crucial. Nevertheless, an appreciation of the British sources of influence goes a long way in explaining the peculiar disciplinary divisions within the Academy and the debate over the status of metaphysics therein. Once we demarcate the critique of systems from the status of the discipline of metaphysics, the establishment of the class of speculative philosophy and the status of metaphysics within it are naturally explained.

The classes up to 1744

The model for dividing the activities of the Berlin Academy into classes may have been the semi-formal structure of the Academy of sciences in Paris. From its beginnings in the mid-1660s the Parisian academy informally divided its subject matter into the mathematical sciences of geometry, astronomy and mechanics, and the physical sciences of chemistry, botany and anatomy.[2] This division was reinforced in the statutes drawn up by Jean-Paul Bignon for the Academy in 1699.[3] The Berlin Academy was first structured in 1710 to comprise four classes, the first two of which were similar to those of the Parisian Academy. They were natural philosophy, including medicine, chemistry and so on; mathematics, including astronomy and mechanics; German language and history, ecclesiastical rather than civil and ancient history; and oriental literature and work for the propagation of the gospel 'among the non-Christian nations'.[4]

2. J. E. McClellan, 'The Académie royale des sciences, 1699-1793: a statistical portrait', *Isis* 72:4 (1981), p.541-67 (543-44).
3. Bernard Le Bovier de Fontenelle, *Histoire du renouvellement de l'Académie royale des sciences en MDCXCIX* (Paris, Jean Boudot, 1708), p.41-42.
4. *Histoire de l'Académie royale des sciences et belles lettres, depuis son origine jusqu'à présent, avec les pieces originales* (Berlin, Haude and Spener, 1750), p.31: 'chez les Nations infideles'. Interestingly, the summary of the classes of 1710 given in the *Histoire de l'Académie royale des sciences et belles lettres* (1746; 1748), p.1-2, differs from the

In the early 1740s, with the reinvigoration of the Academy under the new monarch Frederick the Great, the classes were reshaped significantly. Of the original four classes only mathematics remained unchanged. The classes in January 1744 were the following: natural philosophy (*physique*), mathematics (*mathématique*), philosophy (*philosophie*) and philology (*philologie*). Medicine was omitted, and the study of language and antiquities was expanded to philology, that is, the full range of belles-lettres, by dropping the focus on German and oriental antiquities. This is clear from the published version of the rules of the Academy of 1744 that appears among the various items documenting the history of the Academy in the 1750 edition of the *Histoire*.[5] The preface of the first edition of the *Histoire* (1746), probably written by Samuel Formey,[6] who in 1748 became the perpetual secretary of the Academy, tells us that the new classes represent the combined subjects of two academies, now brought under the auspices of the revivified Académie royale des sciences et belles-lettres: 'The four classes, which compose the Academy of Berlin, embrace all the subjects which elsewhere are the object of two distinct academies, one of the sciences and one of literature.'[7]

Then follows a justification for the inclusion of the class of philosophy, which is worth citing *in extenso*:

> But besides this, we find in our Memoirs the examination of certain questions that are not normally found in similar works, namely, questions of metaphysics, of natural law and of morals. It is for them that the class named philosophy was established. [...]
>
> Metaphysics is without doubt the mother of the other sciences, the theory that provides the most general principles, the source of

original rules set out by Frederick III, elector of Brandenburg. It has 1. natural philosophy and medicine; 2. mathematics; 3. language and German antiquities; 4. languages and oriental antiquities. Perhaps Formey, assuming he as secretary of the Academy was the author, omitted an explicit reference to 'German language' because of the recent decision to commit the Academy to the French tongue; see [Jean Henri Samuel Formey], 'Préface', in *Histoire de l'Académie royale des sciences et des belles lettres de Berlin* (1745; Berlin, Haude, 1746), sig.)( )(1*r-v*.

5. See *Histoire* (1750), p.67. This issue of the *Histoire* differs from those preceding it in the series in so far as it is a full history of the Academy 'from its origins to the present' ('depuis son origine jusquà present', as it says on the title page). This issue also includes a German translation of the documents; see *Histoire* (1750), p.222-23. See also *Histoire* (1746), which lists the old class names in the 'Memoires' section of its contents page and in the 'Memoires' themselves, sig.)( )(4*r-v* and sig.)( )(2*v*.
6. See Terrall, *The Man who flattened the earth*, p.244-45.
7. *Histoire* (1746), sig.)( )(1*v*. All translations are my own unless otherwise stated.

the evidence and the foundation of the certainty of our knowledge. These fine features did not apply, in truth, to the metaphysics of the scholastics, a barren land, which really barely produced briars and thorns. [...]

Great geniuses, by recultivating this portion of the empire of the sciences, have made it assume a completely different aspect. Instead of a dictionary of barbarous terms, we are beginning to have a nursery where each science finds, so to speak, its seeds, and from which are born all the principles, all the guiding notions that steer us whichever way we turn.

Furthermore, the examination of these subjects demands minds free from the shackles of superstitious attitudes that reign in many countries, where the boundaries of reason and faith have not been set in a sufficiently accurate manner. In this regard, we find ourselves in the most favourable situation that one could wish for.

It was, therefore, only natural to eagerly seize hold of these happy beginnings and to work in such a manner as to polish and perfect the keys which unlock all that can be opened by human intelligence. This is the aim of the class of philosophy; it is the object of the new type of memoirs which we are justifying here.[8]

Note here the sense of novelty that accompanies the inclusion of metaphysics within the ambit of the Academy and the very Aristotelian claim that it is the mother of all the other sciences. Note, too, the agricultural metaphors of cultivation and seeds[9] and the overt commitment to a theory of principles. The principles of metaphysics, it is claimed, are those fertile principles that give birth to the principles of all the other sciences. The newly refurbished Academy certainly saw itself as unique, both in its inclusion of the belles-lettres[10] and in its commitment to metaphysics. Yet more change was afoot.

8. *Histoire* (1746), sig)( )(1*v*-)( )(2*v*.
9. Of course, the tree of knowledge was a common metaphor, deployed most famously by Descartes in the dedicatory epistle of the French edition of his *Principles of philosophy*; see René Descartes, *The Philosophical writings of Descartes*, vol.1, ed. John Cottingham, Robert Stoothoff and Dugald Murdoch (Cambridge, 1985), p.186.
10. There is evidence, however, that some people in France believed that the Royal Society of London included belles-lettres within its remit. See the announcement of a French translation of forty volumes of the *Philosophical transactions*, which speaks of it containing not only research on mathematics, natural philosophy, natural history and medicine, but also 'very curious dissertations on belles-lettres', in *Mercure de France* (October 1738), p.2202-2203.

The new classes of 1746

Important, indeed decisive, changes next occurred just eighteen months later in June 1746. These changes followed immediately upon the arrival in Berlin of Maupertuis who was appointed as president of the Academy on 1 February 1746. The register book of the Academy sets out the new rules under which it will operate, the 'Rules for the royal Academy of sciences and belles-lettres' ('Réglement pour l'Académie royale des sciences et des belles lettres'), and these include a description of the four classes, three of which have new names.

> The king [...] wanting to give to this company a final form more appropriate to augment his lustre and its progress, his majesty ordered that from now on it observe the following rules:
>
> I. The Academy will continue as it is, divided into four classes
>
> 1. The class of experimental philosophy will include chemistry, anatomy, botany and all the sciences which are founded on experiment.
> 2. The class of mathematics will include geometry, algebra, mechanics, astronomy and all the sciences which have as their object abstract extension and number.
> 3. The class of speculative philosophy will apply to logic, metaphysics and morals.
> 4. The class of belles-lettres will include antiquities, history and languages.[11]

The class of natural philosophy (*physique*) is now called 'experimental philosophy'. It was to include chemistry, botany and all the sciences founded on experience. The class of philosophy was renamed speculative philosophy and was to include logic, metaphysics and morals. And philology was henceforth to be called belles-lettres and to include the study of antiquities, history and languages. It seems that these new appellations took some time to catch on, for the register book for the very next week mentions the class of philosophy rather than speculative philosophy: 'Professor Formey presented three topics for the prize in the class of philosophy for the following year.'[12] Nevertheless, before long they became standard in the *Histoire* of the

11. See https://digilib.bbaw.de/digitallibrary/digilib.html?fn=/silo10/ARCHIV/I_IV_31/01/&pn=4 (last accessed 14 April 2022).
12. 'Le Professeur Formey ayant presenté trois Sujets indiqués par la Classe de Philosophie pour le Prix de l'année prochaine', https://akademieregistres.bbaw.de/data/protokolle/0002-1746_06_09.xml (last accessed 14 April 2022).

Academy and the new rules with the renamed classes appeared in the *Histoire* of 1748, which details the meetings and work of the Academy from the year 1746.[13]

The dissemination of experimental philosophy

At first sight it may seem that the renaming of the classes in 1746 is mere window dressing. Yet if we step back and take a broader view of developments within the sciences in mid-eighteenth-century Europe, it becomes clear that these changes were far from trivial. The terms 'experimental philosophy' and 'speculative philosophy' had been terms of art in English natural philosophy since the 1660s. They had fairly precise meanings and an important and fascinating history, yet they were only appropriated fully in francophone natural philosophy from the mid-1730s. The story of the dissemination of experimental and speculative philosophy is of great importance for an understanding of the appearance of these terms as titles of classes within the Berlin Academy, and it is worth giving a thumbnail sketch of this history here.

In the late Renaissance it was standard practice to distinguish between the speculative and practical sciences. This distinction had a long pedigree, going all the way back to Aristotle, and it was the most widespread, though not the only, division of the sciences in the early modern period. The speculative sciences were mathematics, natural philosophy and metaphysics, and the practical sciences were ethics, politics and œconomy (see Figure 1).

With the development of the application of experimental methods to natural philosophy in the mid-seventeenth century, a new form of natural philosophical methodology arose in the late 1650s in England, particularly in the work of Robert Boyle. In short, natural philosophy ceased to be regarded as an exclusively speculative science, and a practical form called experimental natural philosophy arose. Once the Royal Society of London was established in 1660, experimental philosophy began to develop and flourish. Books soon appeared such as Boyle's *Usefulness of experimental natural philosophy* (1663) and Henry

13. *Histoire* (1748), p.3-8. Each class was to have a permanent director and up to twelve ordinary academicians as members, and it appears that, apart from experimental philosophy, it was not always easy to find enough suitably qualified savants to fill the classes. The lists of members of each class and external associate members in the year 1750 can be found in the *Histoire* published in that year. See *Histoire* (1750), p.248-58.

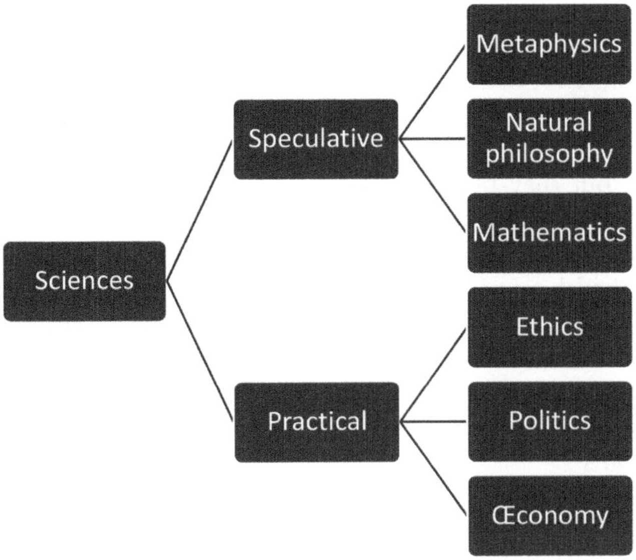

Figure 1: The early modern division of the sciences deriving from Aristotle.

Power's *Experimental philosophy* (1664).[14] When the Royal Society came under attack for practising this new form of natural philosophy, apologists such as Thomas Sprat and Joseph Glanvill came to its defence:

> *Experimental Philosophy* will prevent mens spending the strength of their thoughts about *Disputes*, by turning them to Works [...] And indeed of the usual titles by which men of business are wont to be distinguish'd, the *Crafty*, the *Formal*, and the *Prudent*; [...] The *Formal* man may be compar'd to the meer *Speculative Philosopher*: For he vainly reduces every thing to grave and solemn general *Rules* [...] the *Prudent* man is like him who proceeds on a constant and solid cours of *Experiments*.[15]

Experimental natural philosophy, it was claimed, was superior to its rival speculative philosophy in so far as it gave priority to

14. Robert Boyle, *Usefulness of experimental natural philosophy* (Oxford, Richard Davis, 1663); Henry Power, *Experimental philosophy* (London, John Martin and James Allestry, 1664).
15. Thomas Sprat, *The History of the Royal-Society of London* (London, J. Martyn, 1667), p.341.

observation and experiment over premature system building and the adoption of untested hypotheses without recourse to experiment and observation. In short, experimental philosophers and their defenders defined the new natural philosophy in contrast to speculative philosophy. They were aware of the need, indeed the ultimate goal, of theory, but they believed that a true philosophy of nature could not be achieved until the experimental and observational stage was nearing completion. And in the latter decades of the seventeenth century, British experimental philosophers argued that this could only be achieved through the practice of Baconian natural history. This form of experimental natural history was not the description and classification of natural kinds, but rather an architectonic project encompassing all natural phenomena from the heavens to the bottom of the oceans.[16]

It was not long before experimental philosophy was in the ascendency. The situation is well summed up in John Dunton's student manual, *The Young-students-library* of 1692:

> Philosophy may be consider'd under these two Heads, Natural and Moral: The first of which, by Reason of the strange Alterations that have been made in it; may be again Subdivided into *Speculative* and *Experimental* [...]
>
> But, first, we must consider, the distinction we have made of *Speculative* and *Experimental*, and, as much as possible, Exclude the first, for an indefatigable and laborious Search into Natural Experiments, they being only the Certain, Sure Method to gather a true Body of Philosophy; for the Antient Way of clapping up an entire building of Sciences, upon pure Contemplation, may make indeed an *Admirable Fabrick*, but the Materials are such as can promise no lasting one.[17]

It is clear that throughout the final decades of the seventeenth century in England and in various places on the Continent, such as Italy, the distinction between experimental and speculative philosophy provided the basic terms of reference for the practice

16. For the origins of experimental philosophy, see Peter R. Anstey and Alberto Vanzo, 'The origins of early modern experimental philosophy', *Intellectual history review* 22:4 (2012), p.499-518. For the nature, scope and practice of Baconian natural history, see Peter R. Anstey and and Dana Jalobeanu, 'Experimental natural history', in *The Cambridge history of the philosophy of the scientific revolution*, ed. David M. Miller and Dana Jalobeanu (Cambridge, 2022), p.222-37.
17. John Dunton, *The Young-students-library* (London, John Dunton, 1692), p.vi-vii.

of and theorising about natural philosophy. When in 1713 Isaac Newton publicly declared himself as an experimental philosopher in the second edition of the *Principia* and openly opposed the speculative philosophy of Descartes and Leibniz,[18] the course of experimental philosophy was set: it was now completely dominant. The situation was summed up by Ephraim Chambers in his *Cyclopaedia* of 1728:

> Experimental Philosophy [...] is that which proceeds on *Experiments*; or which deduces the Laws of Nature, and the Properties and Powers of Bodies, and their Actions upon each other, from sensible *Experiments* and Observations [...]
>
> In Effect, *Experiments*, within these 50 or 60 Years, are come into such Vogue, that nothing will pass in Philosophy, but what is founded on *Experiment*, or confirm'd by *Experiment*, &c. So that the new Philosophy is almost altogether *Experimental*.[19]

By this time the Chair of Astronomy and Experimental Philosophy – first held by Roger Cotes – the editor of the second edition of Newton's *Principia* – had been established at Cambridge for over twenty years, and experimental pedagogy was firmly established in Britain with lecture series provided by the likes of Francis Hauksbee the Elder and the Huguenot refugee John Theophilus Desaguliers.[20] In the second decade of the eighteenth century it was embraced in the Netherlands in the writings of Willem 's Gravesande and later Pieter Musschenbroek. Surprisingly, however, it had not held sway in France.

18. Sir Isaac Newton, *The Principia: mathematical principles of natural philosophy*, translated by I. Bernard Cohen and Anne Whitman (Berkeley, CA, 1999), p.943. See also Roger Cotes' 'Preface', in *The Principia*, p.385-99 (386, 398). For the charge that Leibniz was a hypothetical philosopher, see Newton, 'An account of the book entituled *Commercium epistolicum collinii & aliorum*', *Philosophical transactions* 29 (1714-1716), p.173-224 (224).
19. Ephraim Chambers, *Cyclopaedia* (London, James and John Knapton, 1728), p.368.
20. See, for example, John Theophilus Desaguliers' early lecture series *Physico-mechanical lectures, or an Account of what is explain'd and demonstrated in the course of mechanical and experimental philosophy given by J. T. Desaguliers* (London, Richard Bridger, 1717), and Desaguliers, *Lectures of experimental philosophy* (London, W. Mears, 1719). For a detailed treatment of Desaguliers' and Hauksbee's experimental philosophy lectures, see Peter R. Anstey, 'Experimental pedagogy and the eclipse of Robert Boyle in England', *Intellectual history review* 25:1 (2015), p.115-31.

All this was to change in the mid-1730s, such that by 1739 it could be said that experimental philosophy was *à la mode* in France.[21] Three events herald the arrival of experimental philosophy in France. First, Voltaire's *Lettres philosophiques* explicitly endorsed experimental philosophy, tracing its origins back to Bacon: 'In a word, no one before Bacon had known of experimental philosophy.'[22]

Second, the comte de Buffon's preface to his translation of Stephen Hales' *Vegetable staticks* in 1735 shows that he had imbibed the new methodology of the experimental philosophers, and this was later borne out in the numerous Baconian borrowings in the method of natural history in his monumental *Histoire naturelle.*[23] And third, the abbé Nollet, who in the early 1730s had visited Desaguliers in London and 's Gravesande in Holland, returned to Paris and began lecturing on experimental philosophy.[24] He was later appointed to the first French Chair of Experimental Philosophy at the Collège de Navarre in 1753. Meanwhile, Jean D'Alembert, after the publication of the first edition of his *Traité de dynamique* in 1743 began extensive reading in English experimental philosophy, not least the work of James Bradley the discoverer of nutation and lecturer in experimental philosophy at the University of Oxford.[25] Little wonder then

21. See the review of the abbé Pluche's *Spectacle de la nature*, in *Observations sur les écrits modernes, tome dix-septième*, ed. Pierre-François Guyot Desfontaines and François Granet (Paris, Chaubert, 1739), p.101.
22. Voltaire, *Lettres philosophiques* (Amsterdam, E. Lucas, 1734), p.97: 'En un mot personne avant le Chancelier Bacon n'avoit connu la philosophie expérimentale.'
23. See Peter R. Anstey, 'Bacon, experimental philosophy and French Enlightenment natural history', in *Natural history in early modern France: the poetics of an epistemic genre*, ed. Raphaële Garrod and Paul J. Smith (Leiden, 2018), p.205-40.
24. See Jean-Antoine (abbé) Nollet, *Cours de physique expérimentale chez M. l'abbé Nollet* (Paris, P. G. Le Mercier, 1735), and *Programme, ou Idée générale d'un cours de physique expérimentale* (Paris, P. G. Le Mercier, 1738). Maupertuis made a short trip to London in 1728 where he met, *inter alia*, Desaguliers and James Bradley. On his return to France he continued his important experimental work in natural history, but seems not to have openly endorsed experimental philosophy in the manner of Nollet. See, for example, Maupertuis's 'Expériences sur les scorpions', in *Memoires de l'Académie royale des sciences* (Paris, De l'imprimerie royale, 1733), p.223-29, describing experiments performed in 1731. For Maupertuis's London visit, see David Beeson, *Maupertuis: an intellectual biography*, *SVEC* 299 (1992), p.64-66, and Terrall, *The Man who flattened the earth*, p.41-43.
25. Jean D'Alembert, *Traité de dynamique* (Paris, David, 1743). See Peter R. Anstey, 'D'Alembert, the "Preliminary discourse" and experimental philosophy', *Intellectual history review* 24:4 (2014), p.495-516 (508-509).

that experimental philosophy features prominently in his *Preliminary discourse* to the *Encyclopédie* of 1751. Interestingly, the *Dictionnaire philosophique, ou Introduction à la connoissance de l'homme* of 1751 defines *physique* as follows: 'Natural philosophy is the knowledge of the causes and effects of nature. It is experimental or conjectural. Experimental natural philosophy is certain knowledge; conjectural natural philosophy is often only ingenious. The one conducts us to the truth, the other leads to error.'[26]

Thus, when the Berlin Academy adopted the new designations 'experimental philosophy' and 'speculative philosophy' it was continuing the new trend in natural philosophical circles in France, which had its origin in the Royal Society of London.[27] Who the instigators of this change were is yet to be firmly established, but most indicators point to Frederick himself. It seems that the second more specific plan for reform of the Academy that was implemented in June 1746 was very much his own.

In some respects, the Academy's conception of experimental philosophy was the traditional one: it gave an integral place to natural history, and it was opposed to the spirit of systems.[28] Even before the rebranding in 1746 the Academy had included natural history in its first class.[29] And, as in France, the Academy was opposed to systems, enthusiastically embracing Condillac's *Traité des systèmes* in 1749 and offering him foreign membership.[30] Nevertheless, there are two features of the rebranding of the first and third classes that deviate importantly from the situation elsewhere. The first is the absence of any antagonism between experimental and speculative philosophy, and the second the endorsement of the primacy of metaphysics.

26. *Dictionnaire philosophique, ou Introduction à la connoissance de l'homme* (London [i.e. Paris], n.n., 1751), p.261: 'La Physique est la connoissance des causes & des effets de la nature: elle est expérimentale, ou conjecturale. La Physique expérimentale est une connaissance certaine; la Physique conjecturale n'est souvent qu'ingénieuse; l'une nous conduit à la vérité, & l'autre nous mene à l'erreur.'
27. See Leduc, 'Speculative philosophy at the Berlin Academy', p.186.
28. *Histoire* (1750), p.118.
29. For example, see the essay on the natural history of marine fossils by August F. W. Sack, 'Sur de nouvelles pétrifications marines', in *Histoire* (1746), p.67-73, that reflects a form of Baconian natural history, rather than classificatory natural history.
30. Etienne Bonnot de Condillac, *Traité des systèmes* (The Hague, Neaulme, 1749). Condillac was elected as one of the 'Associés Externes' on 4 December 1749: https://akademieregistres.bbaw.de/data/protokolle/0153-1749_12_04.xml (last accessed 14 April 2022).

From its inception, experimental philosophy had been cast as opposed to speculation that paid little heed to experiment and observation, and much of the rhetoric of the experimental philosophers decried the use of hypotheses and pure speculation. The *locus classicus* of English experimental philosophers' negative attitude to hypotheses is, of course, Newton's *hypotheses non fingo*,[31] but it is important to acknowledge that Newton's comment echoes a long-standing and widespread anti-hypotheticalism in British natural philosophy. And this sentiment continued up to the end of the eighteenth century.[32] Thus, we find James Jurin, the secretary of the Royal Society, claiming, 'Sir Isaac Newton has taught us, that the right & indeed the only way to advance real knowledge, is to lay aside all Hypotheses, & to reason only from matter of fact',[33] a sentiment soon repeated by Voltaire in his *Traité de métaphysique*: 'It is clear that one must never make hypotheses.'[34]

In Berlin, by contrast, speculative philosophy was positively encouraged and metaphysics actively pursued. Thus, while there was an institutionalised experimental–speculative distinction in the Berlin Academy, the polarisation of the two disciplines was almost entirely absent. Equally important, however, is the apparent concomitant to the active promotion of speculative philosophy in the Academy, namely, the endorsement of the importance and benefits of the study of metaphysics.

Speculative philosophy and the spirit of systems

According to the standard anglophone historiography of the Enlightenment, this elevated role for metaphysics ran against the dominant trend amongst Enlightenment *philosophes* who, on the whole, decried metaphysics. The Berlin Academy was, on this view, something of an anomaly and, as a result, it was incumbent on its members to defend the prestige that they now accorded to metaphysics. After all, what to many earlier generations of natural philosophers had been almost

31. Newton, *The Principia*, p.943.
32. See Peter R. Anstey, *John Locke and natural philosophy* (Oxford, 2011), p.74-76.
33. Jurin to Thomas Dereham (7 August 1727), in *The Correspondence of James Jurin (1684-1750): physician and secretary to the Royal Society*, ed. Andrea A. Rusnock (Amsterdam, 1996), p.366.
34. Voltaire, *Traité de métaphysique* (1734), in *Œuvres complètes de Voltaire*, vol.22 (Paris, 1879), p.203: 'Il est clair qu'il ne faut jamais faire d'hypothèse.'

anathema, the study of metaphysics, in Berlin was being positively encouraged.[35]

The long-standing historiography that claims that Enlightenment philosophers were deeply anti-metaphysical was endorsed in Peter Gay's influential study *The Enlightenment: an interpretation*. Building on Ernst Cassirer's identification of Enlightenment philosophy with criticism, Gay claimed in 1966 that this led to the raising of 'serious doubts about the value of metaphysics'.[36] Before we analyse the content of Gay's claim here, it is worth pointing out just how widespread this view has been and remains today. Stuart Brown, for example, in his introduction to *British philosophy and the age of Enlightenment* of 1996, speaks of the 'anti-metaphysical tendency' of some Enlightenment philosophers, such as Condillac and Hume,[37] and Nicholas Capaldi speaks of the 'attemped delegitimation of fundamental metaphysical issues' as being unique to the Enlightenment.[38] More recently, William Bristow in his 2017 article on 'Enlightenment' in the *Stanford encyclopedia of philosophy* claims, 'While there are significant Enlightenment thinkers who are metaphysicians – again, one thinks of Christian Wolff – the general thrust of Enlightenment thought is anti-metaphysical.'[39]

Turning to Gay's view, a careful reading indicates that he conflates the widespread anti-system sentiment among the *philosophes* with a rejection of metaphysics. With regard to Condillac, for example, Gay speaks of his refutation of the 'mania for systems' by a form of psychological analysis. He then segues into the claim

35. For Newton's opposition to metaphysics, see Dmitri Levitin, 'Newton and scholastic philosophy', *British journal for the history of science* 49:1 (2016), p.53-77 (67-77).
36. Peter Gay, *The Enlightenment: an interpretation – the rise of modern paganism* (New York, 1966), p.132.
37. Stuart Brown, *British philosophy and the age of Enlightenment* (London, 1996), p.4. See also S. Brown, 'Leibniz and the fashion for systems and hypotheses', in *Philosophers of the Enlightenment*, ed. P. Gilmour (Edinburgh, 1989), p.8-30 (24): 'By the mid-eighteenth century Bayle's scepticism about metaphysics had become pervasive in French philosophical circles'; Mark Hulliung, *The Autocritique of Enlightenment: Rousseau and the philosophes* (Cambridge, MA, 1994), p.181: 'all the philosophes denounced speculative metaphysics.'
38. Nicholas Capaldi, *The Enlightenment project in the analytic conversation* (Dordrecht, 1998), p.20.
39. William Bristow, 'Enlightenment', in *Stanford encyclopedia of philosophy* (autumn 2017 edn), ed. Edward N. Zalta, https://plato.stanford.edu/archives/fall2017/entries/enlightenment/ (last accessed 14 April 2022).

that 'Condillac's argument [in his *Treatise on systems*] impressed his contemporaries as compelling; his criticism of metaphysics, as devastating.'[40] An even more striking instance of this conflation, one that predates Gay, is found in W. H. Barber who claims that Condillac's *Traité des systèmes* 'crystallizes and epitomizes the growing French hostility towards metaphysical speculation'.[41] Again, Barber claims that in the *Preliminary discourse*, D'Alembert 'can record the total defeat of metaphysics'[42] giving credit to Condillac, and yet it is not metaphysics that D'Alembert is referring to here but the spirit of systems to which Condillac has 'delivered the death blow'.[43] In fact, D'Alembert goes on to draw a parallelism between the spirit of systems that pertains to natural philosophy (*physique*) and the manner in which metaphysics stands in relation to geometry: 'The spirit of systems is in natural philosophy what metaphysics is in geometry.'[44]

A thorough study of the origins of the anti-system debate in France reveals that the two are not the same: the critique of systems is not to be equated with a critique of metaphysics. An attack on systems was, from the outset, part and parcel of the British experimental philosophers' critique of speculative *natural* philosophy. It was not focused on metaphysics. An anonymous defence of the Royal Society from 1670 sums up the view nicely: 'they are perswaded with the Excellent Lord *Verulam*, and many other judicious Moderns, that hitherto all sorts of Naturalists have been too hasty in establishing Systems of Natural Philosophy, before they had laid a solid and comprehensive Foundation to build upon.'[45] The view of the early experimental philosophers was not that systems were to be ruled out *tout court*, but that a true system of natural philosophy could only be developed by beginning with experiment and observation. For the first generations of experimental natural philosophers, this was to be achieved by

40. Gay, *The Enlightenment: an interpretation*, p.140.
41. W. H. Barber, *Leibniz in France: from Arnauld to Voltaire* (Oxford, 1955), p.155.
42. Barber, *Leibniz in France*, p.157.
43. Jean D'Alembert, *Preliminary discourse to the Encyclopedia of Diderot*, translated by Richard N. Schwab (Chicago, IL, 1995), p.94.
44. D'Alembert, *Preliminary discourse*, p.95.
45. *A Brief vindication of the Royal Society* (London, John Martin, 1670), p.3, underlining added. This writer is drawing on Robert Boyle's warnings against the danger of systems in *Certain physiological essays* (1660), in *The Works of Robert Boyle*, ed. Michael Hunter and Edward B. Davis, 14 vols (London, 1999-2000), vol.2, p.11.

the construction of experimental natural histories. Thus, Henry Oldenburg, the indefatigable secretary of the Royal Society, told Samuel Sorbière in 1664 that the Royal Society aimed 'to compose in time a Natural and Artificial History which will be perfect, and which can provide materials sufficient to build an unshakably firm system of philosophy'.[46] Fontenelle, the perpetual secretary of the Parisian Academy echoed this sentiment in 1699: 'the Academy is only at the stage of gathering an ample store of well-founded observations and facts, which will one day become the basis for a System. For systematic natural philosophy it must refrain from building its edifice until experimental natural philosophy is able to furnish it with the necessary materials.'[47]

The anti-system sentiment was fully entrenched in France by the mid-1730s, and widespread expressions of this attitude in France appear to have coincided with the acceptance of experimental philosophy there. Buffon, for example, admonishes his readers in 1735, 'let us constantly collect experiments, and flee, if it is possible, from the system-building mentality.'[48] To repeat, it is crucial to note that the opposition to systems is an opposition not to speculative metaphysics, but to the development of sciences in any domain, particularly natural philosophy, without prior recourse to experiment and observation.[49]

46. Oldenburg to Sorbière (3 January 1664, new style), in *Correspondence of Henry Oldenburg*, ed. A. Rupert Hall and Marie Boas Hall, 13 vols (Madison, WI, 1965-1986), vol.2, p.142-44. See also, for example, Oldenburg to Sir John Finch (10 April 1666), in *Correspondence of Henry Oldenburg*, vol.3, p.87.
47. Fontenelle, *Histoire du renouvellement de l'Académie royale des sciences*, sig. ciij*r-v*, underlining added, translation adapted from Roger Hahn, *The Anatomy of a scientific institution: the Paris Academy of sciences, 1666-1803* (Berkeley, CA, 1971), p.33.
48. Comte de Buffon, 'Preface' to *La Statique des vegetaux, et l'analyse de l'air* by Stephen Hales, in *From natural history to the history of nature*, translated by John Lyon and Phillip Sloan (Notre Dame, IN, 1980), p.35-40 (38). For Buffon against systems of classification 'founded on arbitrary principles', see 'Initial discourse', in *From natural history to the history of nature*, p.100-103.
49. For a more detailed treatment of the prehistory of the French anti-system debate, see Peter R. Anstey, 'The principled Enlightenment: Condillac, D'Alembert and principle minimalism', in *Rethinking the Enlightenment: between history, philosophy, and politics*, ed. Geoff Boucher and Henry Martyn Lloyd (Lanham, MD, 2018), p.131-50 (135-39).

The theory of principles and metaphysics

This can be better understood in the light of the theory of how we acquire knowledge in the sciences originating in Aristotle's *Posterior analytics* that was virtually ubiquitous in the eighteenth century. It had three constituents: a theory of principles, a theory of demonstration by which knowledge is derived from those principles, and a theory of the sciences. And it had a prime exemplar, namely, Euclidean geometry. Put simply, most philosophers were committed to the view that in order to have a science, that is, a structured set of propositions analogous to the geometry set out in Euclid's *Elements*, one needed to commence with a determinate set of principles. The question was: how do we get epistemic access to the principles of a science? The experimental philosophers claimed that in the first instance this is via experiment and observation, and not through hypotheses that are posited without recourse to experience. Only then will we be in a position to establish a true system of knowledge on properly established principles.[50] So, the opposition to systems is, for the most part, opposition to *natural philosophical theories* founded on principles that are not based on observation and experiment. What then is the epistemic status of metaphysics in relation to the critique of systems?

In France, Condillac and D'Alembert both openly endorsed metaphysics and worked with a nuanced distinction between particular and general metaphysics, but even here the connection with natural philosophy is evident. Particular metaphysics was the study of the understanding, a form of proto-psychology, and general metaphysics was the theory of being in general. D'Alembert claimed Locke was the champion of particular metaphysics: 'he reduced metaphysics to what it really ought to be: the experimental natural philosophy [*physique expérimentale*] of the soul.'[51] And Condillac's *Essay on the origin of human knowledge* (1746) should be seen as a continuation of Locke's enterprise. The work begins with a measured *apologia* for metaphysics, one that is acutely aware of its dangers and the need for particular metaphysics to be grounded on observation.[52]

50. For further discussion and examples of the application of the theory of principles, see *The Idea of principles in early modern thought*, ed. Peter R. Anstey (New York, 2017).
51. D'Alembert, *Preliminary discourse*, p.82.
52. Etienne Bonnot de Condillac, *Essay on the origin of human knowledge*, translated by Hans Aarsleff (Cambridge, 2001), p.3-5. See also Laurence L. Bongie, who calls the opening paragraph of the *Essay* 'essentially a defence of metaphysics':

Condillac's second book, the *Treatise on systems* (1749), in one sense broadens the anti-system debate. In what is essentially a work expounding a theory of principles, Condillac distinguishes between three types of principles: abstract, speculative and those based on facts. He critiques the first two types by attacking the view, following Locke, that there are innate abstract principles, and then turns to the speculative principles of Spinoza and the monads of Leibniz. He plumps for the third type of principles: 'we rarely take only well-established facts for principles [...] True systems, the only ones that merit the name, are based on principles of this kind.'[53] Yet in doing so, he connects his treatment of systems up with its polemical origins, for his example of a well-established principle is from natural philosophy, namely, Newtonian gravity.

Furthermore, in his *Treatise on systems* Condillac continues to work with the distinction between particular metaphysics and general metaphysics. Thus, his critique of Malebranche pertains to Malebranche's theory of the understanding and the will (chapter 7), that is, particular metaphysics, the very enterprise that Condillac himself had undertaken in the *Essay on the origin of human knowledge* of 1746. One can hardly ignore, then, the parallelism between his attack on Malebranche and his *Essay on the origin of human knowledge* on the one hand, and his attack on Leibniz's monads (chapter 8) and his own *Les Monades*, written *c.*1746, on the other. Little wonder then that he explicitly endorses the value of metaphysical analysis towards the end of the book: 'Metaphysical analysis has the advantage of never ceasing to enlighten the mind: and this because it always makes the mind operate on ideas and which obliges it to follow their generation in a manner so sensible that the mind cannot lose sight of this generation. Thus, it discovers no truth that it does not demonstrate.'[54]

Far from the critique of systems, which found its highwater mark in the *Treatise on systems*, implying an opposition to metaphysics, *pace* Peter Gay and his ilk, Condillac and D'Alembert, along with key members of the Berlin Academy, were, in general, positively disposed

---

'Introduction' to Etienne Bonnot de Condillac, *Les Monades*, ed. Laurence L. Bongie, *SVEC* 187 (1980), p.19.

53. Etienne Bonnot de Condillac, *Treatise on systems*, in *Philosophical writings of Etienne Bonnot de Condillac*, translated by Franklin Philip (Hillsdale, NJ, 1982), p.3. See also *Treatise on systems*, p.123.

54. Condillac, *Treatise on systems*, p.447.

to it. Moreover, when it came to general metaphysics, that is, the science of being, it was standard for proponents of the theory of principles to claim that, over and above those principles that pertain to a particular science and that need to be established via experiment and observation, there were 'common principles' that were known *a priori* or were self-evident.

By the 1740s in Europe many philosophers and natural philosophers had come to believe a cluster of additional theses about the nature and status of principles. These included the view that the sciences are all related on account of their mutual demonstration from some higher-order principles, and that there may be a small number, or even just one, fecund principle or principles from which all the sciences derive. Elsewhere I have called this principle minimalism.[55] Condillac is a case in point, for the subtitle of the first edition of his *Essay* was, 'A work where we reduce to a single principle all that which concerns human understanding', and he claims in the work actually to have discovered this principle.[56] In fact, Buffon had summed it up as early as 1735 in his preface to Hales' *Vegetable staticks*: 'I admit that nothing would be so desirable, as to establish initially a single principle, by which the universe could then be explained.'[57] D'Alembert's Principle, enunciated in his *Traité de dynamique* (1743), and Maupertuis's Principle of least action, first stated in 1744,[58] provided concrete exemplars of the ideal of principle minimalism in natural philosophy. Moreover, Charles Batteux's *Fine arts reduced to a single principle* (1746) and Condillac's chapter on the role of principles in the arts show just how broad-ranging the application of the theory of principles and the commitment to principle minimalism were.[59]

A number of pressing philosophical questions naturally follow from these claims: what are these meta-principles? How do we get epistemic access to them? Do they constitute the principles of a general science of being, namely, metaphysics? And what is their relation to the laws of nature? Indeed, it is these very questions that motivated the prize

55. See Anstey, 'The principled Enlightenment'.
56. Condillac, *Essay on the origin of human knowledge*, p.5.
57. Buffon, 'Preface' to *La Statique des vegetaux*, p.37.
58. Pierre-Louis Moreau de Maupertuis, 'Accord de différentes loix de la nature qui avoient jusqu'ici paru incompatibles', in *Memoires de l'Académie royale des sciences* (Paris, De l'imprimerie royale, 1748), p.417-26.
59. Charles Batteux, *The Fine arts reduced to a single principle*, translated by James O. Young (Oxford, 2015). See also Condillac, *Treatise on systems*, ch.17, p.148-50.

essay competition at the Berlin Academy in 1758, namely, 'Are the laws of statics and mechanics necessary or contingent?'[60]

To sum up, we have: a widespread anti-system sentiment whose origins are found in the opposition of experimental philosophers to speculative natural philosophy; the practice of a particular metaphysics in the study of human understanding; a commitment to an integrated account of the sciences based upon a theory of principles; a commitment to higher-order meta-principles; an openness to the view that these higher-order principles may constitute, in their own right, a science of being or a general metaphysics; and an aspiration that there might be a cluster or even just one ultimate principle by which all natural phenomena might be explained.

The class of speculative philosophy and metaphysics

With this background in mind we can give an entirely natural explanation of Condillac's entry of *Les Monades* for the prize essay at the Berlin Academy in 1746 and 1747: after critiquing Leibniz's monadology in part I, he set about in part II doing general metaphysics the way it should be done. And, as we might expect, Condillac opens his essay with a clear commitment to the theory of principles:

> The best way to discover the truth or falsehood of a system is to find out the sequence of ideas in which the main parts were found and assembled. By analysing the initial notions from which an author starts and the principles which he forms from them, we can see if he has taken the precaution to establish nothing except well defined ideas.[61]

More importantly for our purposes here, we have a straightforward explanation of the inclusion of metaphysics in the class of speculative philosophy and of Formey's emulation of metaphysics as the mother of the sciences in his preface to the *Histoire* of 1746. Returning to his

60. See William Clark, 'The death of metaphysics in enlightened Prussia', in *The Sciences in enlightened Europe*, ed. William Clark, Jan Golinski and Simon Schaffer (Chicago, IL, 1999), p.423-73 (444-48).

61. Condillac, *Les Monades*, p.111: 'Le meilleur moyen de découvrir le vrai ou le faux d'un système, c'est de rechercher par quelle suite d'idées les principales parties en ont été trouvées et assemblées. En analysant les premières notions d'où part un auteur et les principes qu'il en forme, on voit s'il a eu la précaution de ne rien établir que sur des idées bien déterminées.'

comments in the preface to the *Histoire* of that year (quoted earlier), we find that his claims there resonate with the position set out above:

> Metaphysics is without doubt the mother of the other sciences, the theory that provides the most general principles, the source of the evidence and the foundation of the certainty of our knowledge. These fine features did not apply, in truth, to the metaphysics of the scholastics, a barren land, which really barely produced briars and thorns [...]
>
> Great geniuses, by recultivating this portion of the empire of the sciences, have made it assume a completely different aspect. Instead of a dictionary of barbarous terms, we are beginning to have a nursery where each science finds, so to speak, its seeds, and from which are born all the principles, all the guiding notions that steer us whichever way we turn.

Formey's agricultural and nurturing metaphors which describe metaphysics as the 'mother of the other sciences' and the 'seeds' of each science find diagrammatical expression in Diderot's and D'Alembert's system of knowledge appended to the *Preliminary discourse*.[62] For there in the diagram of the tree of knowledge, 'General Metaphysics, or Ontology, or Science of Being in general' stand above all the sciences derived by the faculty of reason. Thus, for both Formey and his French associates, the unity of the sciences derives from the high-order principles and those principles are the constituents of general metaphysics.

We also have an entirely natural explanation of the nature of the prize question on monads for the years 1746 and 1747, for contestants were asked 'after having proved the monads' to 'deduce from it an intelligible explanation of the principal phenomena of the universe, and in particular of the origin and movement of bodies'.[63] This was not simply an exercise in general metaphysics but was designed to determine whether or not the primary constituents of a natural philosophy could be derived from a monadology. That is to say, it was predicated on a theory of principles and the sciences that connected general metaphysics with natural philosophy.

62. D'Alembert, *Preliminary discourse*, p.144-45.

63. Adolf von Harnack, *Geschichte der Königlich Preussischen Akademie der Wissenschaften zu Berlin*, 3 vols (Berlin, 1900), vol.2, p.305. For background to the prize competition on monads, see Thomas Broman, 'Metaphysics for an enlightened public: the controversy over monads in Germany, 1746-1748', *Isis* 103:1 (2012), p.1-23.

Conclusion

It is clear, then, that the new classes of the Berlin Academy represent the institutional instantiation of a Franco-Prussian vision of the structure and epistemic status of the foundations of the sciences. The class of experimental philosophy encapsulated the predominant natural philosophical methodology of the age. The class of speculative philosophy and metaphysics therein gave institutional expression to a theory of the sciences in general that was predicated on the theory of principles that can, rather ironically, be traced all the way back to Aristotle's *Posterior analytics*. While their titles derive from a contrast class that had previously been antagonistic, by the mid-1740s in Prussia most of these tensions had dissipated and the experimental and speculative could cooperatively coexist.

Maupertuis, above all others, is a kind of exemplar of this new disciplinary arrangement. The French experimental philosopher *par excellence* – the one-time natural historian, critic of systems,[64] the leader of the Lapland expedition and the perpetual president of the Academy, the one to whom Pierre Le Monnier dedicated his translation of Roger Cotes' lectures on experimental philosophy[65] – never published a single article in the memoirs of the class of experimental philosophy, or for that matter the class of mathematics. All of his contributions appeared in the memoirs either for the class of speculative philosophy or for that of belles-lettres.[66] This includes his essay on 'The laws of motion and rest deduced from a metaphysical principle'. This appeared in the speculative philosophy memoirs, and yet it contains his most mature treatment of the Principle of least action and a number of mathematical proofs, all of which was later recycled in his work of natural theology entitled *Essai de cosmologie*.[67]

64. For Maupertuis's critique of systems, see his seventh letter, in *Œuvres de Maupertuis*, 4 vols (Lyon, Bruyset, 1768), vol.2, p.257-61.
65. See Roger Cotes, *Leçons de physique expérimentale*, translated by Pierre Le Monnier (Paris, David, 1742), sig. a iij.
66. For Maupertuis's implicit commitment to general and particular metaphysics, see his discussion of speculative philosophy in 'Des devoirs de l'académicien', in *Histoire* (1750), p.115-25 (120-21) (see the appendix to this volume), and the section on 'Expériences métaphysiques', in *Lettre sur le progrès des sciences*, in *Œuvres de Maupertuis*, vol.2, p.373-431 (426-30).
67. Pierre-Louis Moreau de Maupertuis, 'Les loix du movement et du repos déduites d'un principe métaphysique', in *Histoire de l'Académie royale des sciences et belles lettres* (1746; Berlin, Haude, 1748), p.267-94; *Essai de cosmologie* (Berlin, n.n., 1750).

It is little wonder then that years later, when Diderot drew up a plan for a curriculum appropriate for a university in St Petersburg in Russia in 1775, he would recommend that its classes include one for experimental philosophy and natural history, and another for the first principles of metaphysics.[68]

68. Denis Diderot, *Plan d'une université pour le gouvernement de Russie*, in *Œuvres complètes de Diderot*, ed. Jules Assézat, 7 vols (Paris, 1875-1877), vol.3, p.409-546 (451).

'Mother of all sciences' or mere speculation? The justification of metaphysics at the Berlin Academy between 1746 and 1765

ANNELIE GROSSE

Metaphysics and the renewal of the Berlin Academy

The role of metaphysics at the Royal Academy of Science and Literature in Berlin was tightly linked to the extraordinary establishment of a class of speculative philosophy (*classe de philosophie spéculative*) in 1744/1746.[1] In addition to logic and morals, metaphysics was listed as a core subject of this class, which did not exist as such either in the previous stage of the Berlin Academy or in any other European academy or learned society of the time.[2] It was with this speculative philosophy class that the Academy was able to emphasise

I presented a first version of this chapter at the 'Scientiae' conference in Oxford in 2016. I am grateful to Peter Anstey and Tinca Prunea-Bretonnet for their comments on several versions of this chapter and suggestions for further reading. To Philip Jakes I am grateful for English-language corrections. All translations in this chapter are my own unless otherwise indicated.

1. The dating 1744/1746 refers to the Academy's renewal in two moments: in 1744 the fusion of the old Academy and the Société littéraire was achieved and new statutes released. In 1746, after the eventual arrival of the Academy's new president Maupertuis, this foundation was confirmed and a revised version of the statutes issued. The class of speculative philosophy was part of both statutes. On the particularity of the speculative philosophy class at the Berlin Academy and the innovative concept of speculative philosophy that was established by its members, see Christian Leduc, 'Speculative philosophy at the Berlin Academy', in *History of universities*, vol.33.2, ed. Mordechai Feingold and Andrea Sangiacomo (Oxford, 2020), p.175-200.
2. See 'Réglement de l'Académie, 10. Mai 1746', in Adolf von Harnack, *Geschichte der Königlich Preussischen Akademie der Wissenschaften zu Berlin*, 3 vols (Berlin, 1900), vol.1.1, p.300. Before 1744 the Berlin Academy was divided into the following four classes: mathematics, natural philosophy, German language and history, and oriental literature. See also Mary Terrall, 'The culture of science in Frederick the Great's Berlin', *History of science* 28 (1990), p.333-64 (351-52).

its outstanding character among its European competitors and, as I shall argue, the concept and practice of metaphysics played a key role in the legitimation of this undertaking. However, since metaphysics itself lacked broad support and legitimacy in the scholarly world of the time, the academicians at Berlin had to reinvent the discipline, and they attempted this with some audacity in the opening pages of the renewed Academy's first publication, its 1745 yearbook, issued in 1746.[3] In the 'Préface' to this yearbook, the universal and fundamental character of metaphysics was stressed and at the same time its reformation in Berlin proclaimed:

> But in addition to this, one will find in our *Mémoires* the examination of certain questions that usually find no space in this kind of book. Such are the questions of metaphysics, natural law and morality. It is for them that a class has been founded which is called PHILOSOPHICAL. This sort of uniqueness deserves that we interrupt our considerations for a moment in order to justify it. Metaphysics is indisputably the mother of the other sciences, the theory which provides the most general principles, the source of evidence and the basis of certain knowledge. These fine qualities did not characterize in truth the metaphysics of the scholastics: barren soil that produced hardly anything other than brambles and thorns. And since they did not know any other [form of metaphysics] when the major academies were founded, they excluded it with a sort of disdain, and they considered it as an obstacle rather than an aid to the extension of our knowledge. Great geniuses, by recultivating this section of the empire of sciences, have covered it with a completely different face. Instead of a dictionary of savage terms, we begin to have a nursery in which every science finds, to say it like this, its seeds and from which arise all principles, all guiding notions that lead us in all respects.[4]

3. William Clark in 'The death of metaphysics in enlightened Prussia', in *The Sciences in enlightened Europe*, ed. William Clark, Jan Golinski and Simon Schaffer (Chicago, IL, 1999), p.423-73 (423), describes Prussia's endeavour to save metaphysics from its certain decline as an expression of its provinciality.
4. [Jean Henri Samuel Formey], 'Préface', in *Histoire de l'Académie royale des sciences et des belles lettres de Berlin* (1745; Berlin, Haude, 1746), sig.)()(*v*-)()(2*v*. Translated from French: 'Mais outre cela on trouvera dans nos Mémoires l'examen de certaines Questions, qui n'entrent pas ordinairement dans de semblables Ouvrages. Telles sont les Questions de Metaphysique, de Droit naturel & de Morale. C'est à leur intention qu'a été fondée la Classe qu'on nomme PHILOSOPHIQUE. Cette espece de singularité mérite que nous nous arrêtions un moment à la justifier. La Metaphysique est sans contredit la Mére des autres

This statement informed the European scholarly public that all the other European academies that were founded earlier than the Prussian one had had or still had an incorrect image of metaphysics, in that they still equated it with unsupportable and outdated scholastic system building. Hence, this statement implicitly announced a new era. The character of this new era was defined by an enhanced degree of universality, evidence and certainty of metaphysical knowledge, paired with a strong rejection of scholastic pedantry.

This legitimation of metaphysics as a certain and particularly fundamental science that provided the basis and evidence for all other sciences not only had an impact on the perception of the Berlin Academy in the European learned world – an aspect that I will not address in this chapter – it also had a significant effect on the institution's internal configuration, more precisely on the relation between the four different classes, as well as between the individual members and on the evolution of these relations over the years. As I will argue, the definition of metaphysics, and especially its potential interaction with other sciences or disciplines of knowledge, was under permanent scrutiny in the academic works for at least the twenty years that followed the Academy's renewal. I shall argue that it was the decades after the Academy's renewal of 1744-1746 that constituted the real process of reinventing metaphysics. Unlike the way in which it had been portrayed in the first yearbook, the supremacy of metaphysics at the Academy was far from being a stable and accepted doctrine at the moment when the renewed Academy began to set about its work.

If we consider the structure of the Academy in 1744-1746, however, the actual power relations between the disciplines do not seem to match the ideal of metaphysics' superiority as stated in the first

---

Sciences, la Theorie qui fournit les principes les plus généraux, la source de l'evidence, & le fondement de la certitude de nos connoissances. Ces beaux caractéres ne convenoient pas à la verité à la Metaphysique des Scolastiques, terre ingrate, qui ne produisoit gueres que des ronces et des épines. Et comme on n'en connoissoit point d'autre, lorsque les principales Académies ont été fondées, on l'a laissée à l'ecart avec une espéce de dédain, & on l'a regardée comme un obstacle plutôt que comme une aide à l'etenduë de nos connoissances. De grands Genies, en donnant une nouvelle culture à cette portion de l'Empire des Sciences, lui ont fait revêtir une toute autre face. Au lieu d'un Dictionnaire de termes barbares, nous commençons à avoir une pépiniere, où chaque Science trouve, pour ainsi dire, sa semence, et d'où naissent tous les principes, toutes les notions directrices qui nous guident, de quelque coté que nous tournions nos pas.'

yearbook. First, due to the particular genesis of the new Academy, in which the members of two precursor institutions were integrated, the four classes were very unequal both in their numbers and in the proficiency of their members. Moreover, the numerical strength of a class did not necessarily correspond to its productivity (i.e. its published output). The class of mathematics, for example, had the smallest number of members (four), but it produced the most academic treatises.[5] This high proficiency in mathematics was due to the extraordinary output of one person, the Swiss Leonhard Euler who had the reputation of being the world's finest mathematician and who had joined the Berlin Academy in 1741, well before its definitive renewal.[6] Euler produced thirty academic *mémoires* between 1746 and 1750, and in this was well ahead of his colleagues André Sigismund Margraff of the class of experimental philosophy (ten), Johann Kies in mathematics (nine) and Samuel Formey of the class of speculative philosophy (eight).

This imbalance between the individual members' outputs is reflected in a striking inequality in conceptual power that divided mathematics and physics, on the one hand, from philosophy and literature on the other. During the deliberations on how to organise the new Academy at the beginning of 1744, one of the curators and Frederick's advisor, Samuel von Schmettau, had suggested suspending the salaries of the members of the speculative philosophy and belles-lettres classes until the new institution's finances were more stable. The argument for investing in the natural philosophy class and

5. See 'Liste de l'Académie royale des sciences et des belles lettres de Berlin pour l'année MDCCXLVII', Archives de l'Académie des sciences et de l'Institut de France, Paris, fonds 43 J, Pierre Louis Moreau de Maupertuis, dossier 63. The ordinary members in the four classes are (1) *Classe physique*: Johann Theodor Eller (director); Augustin Buddeus; Pierre Carita; Joseph de Francheville; Johann Gottlieb Gleditsch; Johann Nathanael Lieberkühn; Michael Matthias Ludolff; Christian Friedrich Ludolff; André Sigismund Margraff; Johann Heinrich Pott; August Friedrich Sack; Samuel Scharschmidt; Theodor Sprögel; (2) *Classe mathématique*: Leonhard Euler (director); Augustin Grischau; Abraham Humbert; Johann Kies; (3) *Classe philosophique*: Johann Philipp Heinius (director); Antoine Achard; François Achard; Samuel Formey; Philippe Joseph de Jariges; Timotheus Christian Stubenrauch; (4) *Classe philologique*: marquis d'Argens (director); Jakob Elsner; Georg Karl Hering; Georg Gottfried Kuster; Simon Pelloutier; Johann Peter Süßmilch.
6. See Eduard Winter, *Die Registres der Berliner Akademie der Wissenschaften 1746-1766: Dokumente für das Wirken Leonhard Eulers in Berlin zum 250. Geburtstag* (Berlin, 1957), p.16-19.

the mathematics class instead of in philosophy and literature was that the former had to perform more costly experiments and their research was more useful to the state.[7] The king, who acknowledged these considerations, indeed suspended the payments initially, which added to the perception of the two classes concerned as being under-represented, and seemingly made some of their members feel exploited. This precarious situation of the class of philosophy in the very first years of the renewed Academy seems to be at odds with the perception of metaphysics' leading role in the whole of the Academy's disciplines, but it also explains the apparent need of its practitioners to emphasise its significance in their *mémoires*.

The most ardent advocate for the speculative philosophy class, with respect not only to its very practical institutional recognition but also to its scientific excellence, was the Huguenot pastor and professor of philosophy Jean Henri Samuel Formey.[8] During the initial years of the renewed Academy, he strongly defended the cause of metaphysics as a universal and certain science and reiterated the proclamation of metaphysics' superior role published in the first yearbook. It is highly probable that it was Formey who wrote this proclamation in the yearbook's 'Préface'. Having been appointed as the Academy's historian in July 1745, he was entrusted with the compilation and publication of the yearbook.[9] Moreover, in terms of its content the 'Préface' shows much resemblance to the teachings of the German philosopher Christian Wolff, whose ideas had been at the centre of Formey's work as a theologian and philosopher since the early 1740s.[10] In Wolff's classification of the sciences, ontology, which was

7. See Samuel von Schmettau to Frederick II, 15 January 1744, Archiv der Berlin-Brandenburgischen Akademie der Wissenschaften (BBAW), Berlin, I. Historische Abteilung, Abt. I Nr. 5, f.144*r*. If we believe Harnack, *Geschichte*, vol.1.1, p.289, n.6, then Schmettau initially wanted to dispense entirely with the philosophical and philological classes.
8. On Formey's biography and thought, see Jens Häseler, 'Samuel Formey, pasteur huguenot entre Lumières françaises et Aufklärung', *Dix-huitième siècle* 34 (2002), p.239-47, and Martin Fontius, '"Libertas philosophandi" und "siècle de la philosophie": zum geistesgeschichtlichen Standort Formeys', in *Franzosen in Berlin: über Religion und Aufklärung in Preußen. Studien zum Nachlass des Akademiesekretärs Samuel Formey*, ed. Martin Fontius and Jens Häseler (Basel, 2019), p.125-252.
9. See 'Actum' of 15 July 1745, Archiv der BBAW, I. Historische Abteilung, Abt. IV Nr. 12, f.114.
10. In 1741 the first volume of Formey's *La Belle Wolfienne* appeared, a philosophical novel in which the basics of Wolff's philosophy are abridged and explained for a broad public; see Formey, *La Belle Wolfienne*, 6 vols (The Hague, Le Vier, 1741,

one of metaphysics' three subdisciplines, was assigned the role of *philosophia prima* or *Grundwissenschaft*, for it dealt with the most basic and general notions of being. Therefore, the role of ontology and, by extension, of metaphysics was to explain and to transmit these general notions and principles to all other sciences.[11] In the Academy's statement on metaphysics cited above, this idea corresponds to the expression that metaphysics is the source of evidence and certainty of all knowledge. Wolff particularly underlined the methodological impact of ontology as *philosophia prima*, and praised himself for having reformed the confusing scholastic concept of ontology and having established clear principles, the so-called guiding notions (*notiones directrices* or *notions directrices*) – a term that we find also in the above-cited 'Préface' – which were able to guide human understanding in its search for truth. Wolff had developed his thoughts on ontology and its guiding notions in a Latin treatise of 1729 that Formey translated for the introduction of the fourth volume of his *Belle Wolfienne* (*The Beautiful Wolffian*), which appeared in the same year as the Academy's first yearbook, in 1746.[12]

---

and Neaulme, 1743-1753). In his role as a professor of philosophy at the Collège français in Berlin, Formey also developed a textbook that abridged the entire Latin oeuvre of Wolff: Formey, *Elementa philosophiae seu medulla wolfiana in usum auditorum* (Berlin, Haude and Spener, 1746).

11. See Christian Wolff, 'Vorbericht von der Welt-Weisheit', in *Vernünfftige Gedancken von den Kräften des menschlischen Verstandes und ihrem richtigen Gebrauche in Erkäntniss der Wahrheit*, in *Gesammelte Werke*, 4th edn, section I, vol.1 (Hildesheim, 2006), p.115-22, §14, and *Discursus praeliminaris de philosophia in genere / Einleitende Abhandlung über Philosophie im Allgemeinen. Historisch-kritische Ausgabe*, ed. Günter Gawlick and Lothar Kreimendahl (Stuttgart and Bad Cannstatt, 1996), §73. According to Wolff's classification, the other two subdisciplines of metaphysics are pneumatology and natural theology.

12. See Formey, 'Introduction à la métaphysique, ou dissertation préliminaire sur les notions directrices & sur le véritable usage de l'ontologie', in *La Belle Wolfienne*, vol.4 (The Hague, Neaulme, 1746), p.i-xxiii. The original text from which Formey translated is entitled 'De notionibus directricibus & genuino usu philosophiae primae' and had appeared in Wolff's *Horae subsecivae Marburgenses anni MDCCXXIX, quibus philosophia ad publicam privatamque utilitatem aptatur*, in *Gesammelte Werke*, 1st edn (1729), section II, vol.34.1 (Hildesheim, 1983), p.310-50. Remarks on the connection between this writing by Wolff and Formey's academic work can be found in Stefanie Buchenau, 'Notions directrices et architechtonique de la métaphysique: la critique kantienne de Wolff en 1763', *Astérion* 9 (2011), doi:10.4000/asterion.2136, and Tinca Prunea-Bretonnet, 'La méthode philosophique en question: l'Académie de Berlin et le concours pour l'année 1763', *Philosophiques* 42:1 (2015), p.107-30.

Considering Formey's effort in apprehending and disseminating Wolff's philosophy, it might seem that the emphasis on metaphysics at the renewal of the Berlin Academy formed part of Formey's personal endeavours as a partisan of Christian Wolff, rather than the established self-description of the whole body of the Academy members. This would explain why, in subsequent years, Formey was more or less the only one who defended the superiority of metaphysics by rigorously referring to Wolffian methodology. His immediate opponents in this undertaking were the Academy's president Pierre-Louis Moreau de Maupertuis and the mathematician Leonhard Euler, who were both critical and even polemical about Wolff, as became particularly obvious in the Academy's famous essay contest on the Leibnizian monadology in 1747.[13] Nevertheless, the evidence suggests that the often-claimed antagonism between Newtonian and Wolffian thought at the Berlin Academy[14] evolved to a large extent on a formal level instead of being always directly grounded in philosophical (content-related) differences between two 'schools of thought'. This distinction between formal and actual philosophical antagonism will help us to understand why Formey was able to 'impose' his Wolffian take on metaphysics on the whole Academy in 1746 without any protest from his colleagues, especially from the new president Maupertuis, who was (and still is) considered to be a Newtonian and anti-Wolffian. My contention is that, if we approach the problem of the status of metaphysics in terms of its actual content, the general agreement of the Academy, and even

13. See Winter, *Registres*, p.9, and Clark, 'Death of metaphysics', p.439-40. Instead, Mary Terrall, *The Man who flattened the earth: Maupertuis and the sciences in the Enlightenment* (Chicago, IL, 2002), p.264, concludes that Maupertuis, unlike Euler, refrained from openly attacking Wolff in the debate on monadology. The tension between Euler and Maupertuis on the one hand and Wolff and his followers on the other was not only one-sided: Johannes Bronisch, *Mäzen der Aufklärung: Ernst Christoph von Manteuffel und das Netzwerk des Wolffianismus* (Berlin, 2010), p.197-201, provides a comprehensive account about Wolff's and his supporters' – the so-called Aléthophiles – hostility towards the Berlin Academy's new personnel.
14. See for example Harnack, *Geschichte*, vol.1.1, p.432-33, and Ronald S. Calinger, 'The Newtonian–Wolffian controversy: 1740-1759', *Journal of the history of ideas* 30:3 (1969), p.319-30. Instead, Thomas Ahnert, 'Newtonianism in early Enlightenment Germany, c. 1720 to 1750: metaphysics and the critique of dogmatic philosophy', *Studies in history and philosophy of science* 35:3 (2004), p.471-91 (483-87), argues that anti-Wolffian positions within the Berlin Academy in the mid-eighteenth century only rarely had Newtonian sources.

Maupertuis, with the declaration of metaphysics as *Grundwissenschaft* à la Wolff is entirely natural.

That Maupertuis agreed with Formey's Wolffian-inspired 'Préface' is also confirmed by the records of the Academy, which suggest that the 'Préface' must have gone through Maupertuis's hands before publication. As a note of 22 September 1745 in the Academy's registers shows, on the occasion of Maupertuis's eventual arrival in Berlin, the members of the Academy were ordered to hand over the current statutes and everything that had been written so far for the first volume of the yearbook for the new president to examine.[15] If Maupertuis had not approved the material that the academicians had given to him, which certainly also included Formey's 'Préface', it would not have been published as it was. Therefore, it seems that Maupertuis's and Formey's (Wolffian) conceptions of metaphysics were far closer than the narrative of a Wolffian–Newtonian divide at the Berlin Academy would lead us believe, and, hence, that the new president, at least in the initial years of the renewed Academy, supported the idea of metaphysics' primacy among the disciplines as it had been displayed in the first yearbook. To a certain extent this congruency between Formey's and Maupertuis's opinions on metaphysics' general function is also confirmed by the first discourse Maupertuis presented to the Academy in 1746; yet at the same time this discourse also revealed a slight, but important, divergence between the two philosophers' conceptions of metaphysical principles and method.

Metaphysics' threat from mathematics

Shortly after the renewed Academy had started its work, in 1746, Maupertuis presented a treatise in which he deduced the existence of God from a metaphysical principle. The following year, Formey responded by delivering two consecutive treatises on the same topic, which allowed him to develop further the Wolffian concept of guiding or common notions that had already been alluded to in the 'Préface' of the first yearbook. This debate between Maupertuis and Formey on the existence of God reveals that, while they both generally acknowledged the function of metaphysics as a science of universal principles, they had completely different conceptions as to

15. See 'Actum' of 22 September 1745, Archiv der BBAW, I. Historische Abteilung, Abt. IV Nr. 12, f.119.

how these principles were generated, applied and interacted with other sciences.[16]

Maupertuis opened this debate on 6 October 1746 with a treatise entitled 'Les loix du mouvement et du repos déduites d'un principe métaphysique' ('The laws of motion and rest deduced from a metaphysical principle'), in which he set out to find the most evident and powerful way of proving the existence of God. The purpose of his treatise was to criticise and likewise to reform the approach of physico-theology according to which God's existence could be proved directly through the observation of nature.[17] In Maupertuis's view, instead of proving God through all the different phenomena of His creation, like an insect or a flower, the superior approach was to prove Him through the universal first laws of nature on which these phenomena depended:

> The organisation of the animals, the multitude and smallness of the insects' parts, the vastness of the celestial bodies, their distance and rotations, serve rather to surprise our mind than to enlighten it. The supreme being is everywhere; but He is not visible in the same way everywhere. We will be able to better perceive Him in the most simple objects: let us search for them in the first laws that He has imposed on nature; in these universal rules according to which movement is kept, distributed or destroyed, and not in the phenomena that are nothing other than the too complicated consequences of these laws.[18]

16. On Maupertuis's and Formey's proofs of God's existence, see Tinca Prunea-Bretonnet, 'Prouver l'existence de Dieu à partir de l'observation de la nature: Formey (et Wolff) sur l'argument des causes finales', in *Berliner Aufklärung: Kulturwissenschaftliche Studien*, ed. Ursula Goldenbaum and Alexander Košenina, vol.7 (Hanover, 2020), p.165-83. She underlines the difference between their methodological approaches to the subject and describes Formey's proof as a tacit refutation of Maupertuis's (p.166).
17. See Pierre-Louis Moreau de Maupertuis, 'Les loix du mouvement et du repos déduites d'un principe métaphysique', in *Histoire de l'Académie royale des sciences et belles lettres* (1746; Berlin, Haude, 1748), p.267-97 (269). In 1750 Maupertuis published an extended version of his 'Loix du mouvement' under the title *Essai de cosmologie* (Berlin, n.n., 1750), which became one of his most famous works.
18. Maupertuis, 'Les loix du mouvement', p.278-79: 'L'organisation des Animaux, la multitude & la petitesse des parties des insectes, l'immensité des corps celestes, leurs distances & leurs révolutions, sont plus propres à étonner notre esprit qu'à l'éclairer. L'Etre suprême est par tout; mais il n'est pas par tout également visible. Nous le verrons mieux dans les objets les plus simples: cherchons-le dans les premieres loix qu'il a imposées à la Nature; dans ces regles universelles, selon lesquelles le Mouvement se conserve, se distribue, ou se détruit; & non pas dans les Phenomênes qui ne sont que des suites trop compliquées de ces loix.'

By referring to first and universal laws, as opposed to phenomena, Maupertuis seemed to suggest a metaphysical approach that was precisely in the vein of the 1745 postulate of metaphysics as fundamental science of universal principles. This might have been what Formey himself had in mind when, one year later, in the second of his two treatises, entitled 'Examen de la preuve qu'on tire des fins de la nature, pour établir l'existence de Dieu' ('Examination of the proof of God's existence drawn from the purposes of nature'), he claimed to pursue the same objectives as Maupertuis before him.[19] Indeed, Formey's treatise was also an onslaught on the physico-theological argument from design and, following Maupertuis, he made a plea for tracing all natural phenomena to the most simple and general principles on which they depend.[20]

Like Maupertuis, Formey also operated with the concept of the laws of motion to which – in a first step of generalisation – the phenomena of nature could be reduced. These laws of motion, considered as general principles of nature, were usually established as *a priori* principles through mathematical demonstration, and were confirmed by experience. Yet neither Maupertuis nor Formey judged them reliable and certain enough to deduce the proof of God's wisdom and power from them; in Maupertuis's view, the existing laws of motion did not rely on a correct *a priori* method since they were established on mere hypotheses instead of truly mathematical notions. Therefore, he preferred to deduce the laws of motion rather from the attributes of a supreme being.[21] The principle that corresponded most to God's wisdom and strength, and thus constituted the most certain source of knowledge, was in Maupertuis's view his own Principle of least action, according to which changes in nature were incited by the smallest quantity of power possible.[22] Maupertuis, therefore, strove to establish the laws of motion in a different yet still rational or *a priori* manner. Formey rejected the common laws of motion, not so much for their defective deduction from hypothetical principles, but rather for their inability to account for their own contingent nature (i.e. for the contingent world order), which was an essential prerequisite

19. See Formey, 'Examen de la preuve qu'on tire des fins de la nature, pour établir l'existence de Dieu', in *Histoire de l'Académie royale des sciences et belles lettres* (1747; Berlin, Haude and Spener, 1749), p.365-84 (365).
20. See Formey, 'Examen de la preuve', p.369.
21. See Maupertuis, 'Les loix du mouvement', p.279.
22. See Maupertuis, 'Les loix du mouvement', p.286.

of God's necessary existence. For him, the contingency of the laws of motion could only be assured if they were deduced from the Principle of non-contradiction and the Principle of sufficient reason, the two Leibnizian principles which, in Wolff's ontology, counted as the most basic notions.[23] Although the metaphysical principles to which Maupertuis and Formey referred were different, they emerged out of the two philosophers' shared scepticism towards those *a priori* principles from which the laws of motion had previously been deduced. Moreover, Maupertuis's and Formey's alternative principles had the same function: they were supposed to provide the laws of motion with a more solid and universal foundation, which in turn illustrates both authors' general trust in *a priori* principles and methods.

However, the trust in *a priori* principles that they both seemed to profess was in reality not equally unconditional, which in turn crucially affected their respective concepts of the role of metaphysics. The difference between Maupertuis and Formey was that the former did not consider his *a priori* Principle of least action as uniquely sufficient and self-sustained in the establishment of the laws of motion, whereas the latter was convinced of the stand-alone quality of a simple metaphysical notion like the Principle of sufficient reason. Maupertuis had constructed his proof of God on two sources of knowledge that confirmed each other and thus reinforced the proof: he declared that the laws of motion that he deduced from the Principle of least action were the same as could be induced through the observation of nature. Precisely the compatibility of the results of the two epistemic procedures – deduction and induction – was a sign of the truth of his principle and, as a consequence, of God's existence.[24] In terms of the relation between the disciplines of metaphysics (more precisely metaphysics of nature, i.e. cosmology) and natural philosophy (more precisely mathematical mechanics), Maupertuis's epistemological approach implied their mutual dependence. Christian Leduc has described Maupertuis's conception of a metaphysics of nature, or cosmology, as follows: Maupertuis considered metaphysics as a foundational science, though a science that received its legitimation only on the basis of

23. See Formey, 'Examen de la preuve', p.381, and 'Introduction à la métaphysique', p.vii-ix and xi-xiii. In short, Formey's argument in his 'Examen de la preuve' evolved as follows: a law of nature existed because there was a sufficient reason for its existence (Principle of sufficient reason); yet its existence is contingent since its non-existence was not impossible (Principle of non-contradiction).

24. See Maupertuis, 'Les loix du mouvement', p.279 and 286.

previously established physical laws that were confirmed by experience and mathematical calculation. As such, Leduc qualified Maupertuis's position as being in between the Wolffian (and Kantian) conception of an entirely *a priori* metaphysics as a necessary precondition of physical laws and the conception of D'Alembert and Voltaire who defended the independence of natural philosophy from a metaphysical foundation.[25]

In contrast to Maupertuis, Formey defended the unconditional legitimacy of *a priori* ontological principles to account for the secondary principles and phenomena of all other sciences. This legitimacy was, according to Formey, due to the epistemic status of these *a priori* principles. As Formey explained in his 'Examen de la preuve', Leibniz had declared the Principle of sufficient reason to be an axiom, for it was self-evident and there was no argument that was able to disprove its truth, unless one accepted something could emerge out of nothing.[26] Having clarified this, Formey could also claim with confidence that all the attempts of mathematicians to prove certain laws of motion eventually led back to the most simple ontological principle, that is sufficient reason.[27] This claim was strongly indebted to Wolff's considerations on the role of his ontological basic notions that were supposed to guide every sort of knowledge acquisition in the sense that they constituted the primitive source of the individual principles of each science, even of Euclid's mathematical axioms.[28]

The question of the relation of mathematics to metaphysics was an additional aspect that differentiated Maupertuis's and Formey's opinion on the role of metaphysics. For both philosophers this seems to have been predominantly a methodological question, and we probably have to see it in the context of a debate that had occupied European scholars since the seventeenth century: the debate as to whether mathematical method should be applied in philosophy.[29] Maupertuis, in his treatise of 1746, left no doubt that he entirely rejected attempts to apply mathematical terms and methods to metaphysical subjects. According to him, such attempts only appeared mathematical on

25. See Christian Leduc, 'La métaphysique de la nature à l'Académie de Berlin', *Philosophiques* 42:1 (2015), p.11-30 (15).
26. See Formey, 'Examen de la preuve', p.370.
27. See Formey, 'Examen de la preuve', p.380.
28. See Formey, 'Introduction à la métaphysique', p.xviii-xix.
29. See Giorgio Tonelli, 'La disputa sul metodo matematico nella filosofia della prima meta del settecento e la genesi dello scritto Kantiano sull'evidenza', in *Da Leibniz a Kant: saggi sul pensiero del settecento*, ed. Claudio Cesa (Naples, 1987), p.81-107.

their surface since metaphysical subjects were too complex to be truly demonstrated by mathematics, the science of the simple things. At the same time, however, Maupertuis was convinced that mathematics was able to provide a much more solid and certain proof of God and, therefore, could function as an alternative to metaphysics:

> Let us see if we can make a more fortunate use of this science [mathematics]. The proofs of God's existence that it provides will have the advantage over all others that they possess the evidence that is characteristic of the mathematical truths. Those who do not trust enough in metaphysical reasoning will find more reliability in this sort of proofs [the mathematical ones]; and those who do not esteem popular proofs so much will find in them more accuracy and superiority.[30]

Maupertuis's argument in favour of the extension of mathematics' scope to the realm of questions such as the existence of God relies on his assumption that the mathematical method possesses a much higher degree of certainty than traditional metaphysical methods. His preference for mathematical demonstration is also evident in the last part of his treatise, where he explains the laws of the distribution of motion between two bodies that hit each other: his explanation is entirely mathematical, relying on the calculation of velocity. Nonetheless, he arranges these calculations under the umbrella of his metaphysical Principle of least action.[31] Hence, although Maupertuis, as we have shown, acknowledged the foundational function of metaphysical principles for other sciences, he did not trust its methods and hence backed up the principles with calculation.

Formey's metaphysical considerations on God's existence seem to reply directly to Maupertuis's criticism of the reliability of metaphysical method. In the first of his two treatises of 1747, entitled 'Les preuves de l'existence de Dieu, ramenées aux notions communes' ('The proofs of God's existence traced back to common notions'), Formey acknowledges the similarity between traditional

30. Maupertuis, 'Les loix du mouvement', p.278: 'Voyons, si nous pourrons faire un usage plus heureux de cette science [la mathématique]. Les preuves de l'existence de Dieu qu'elle fournira, auront sur toutes les autres, l'avantage de l'evidence qui caracterise les verités mathématiques. Ceux qui n'ont pas assez de confiance dans les raisonnemens métaphysiques, trouveront plus de sûreté dans ce genre de preuves: & ceux qui ne font pas assez de cas des preuves populaires, trouveront dans celles-ci plus d'exactitude & d'élévation.'
31. See Maupertuis, 'Les loix du mouvement', p.290-94.

metaphysical deduction and the mathematical method,[32] yet he insists that metaphysics is more than just a variety of or even an alternative to mathematics. The argument by which Formey elevates the discipline of metaphysics to a position above mathematics is not linked to method though; instead, he highlights metaphysics' impact on human life, more precisely on the most important questions, like religion and morality:

> If for many centuries the mathematicians have been perceived as the only custodians of certainty, it was because their science was the only one that possessed indisputable principles and that never lost track of them. [...] It is not true that the first notions of metaphysics were less clear and carved with less indelible letters in our minds than those of mathematics. But they are more difficult to discover because they have had for all times and still today have to fight the most dubious enemies of all truth: the senses and the imagination which obfuscate these truths in the majority of people – even in philosophers – and which suspend our agreement. You can add to this that from the mathematical truths result no consequences that affect our heart or our passions, whereas this is not the same with the metaphysical notions, which necessarily influence natural theology and morality.[33]

It was at this point that the question of how to prove the existence of God eventually became, in the hands of Maupertuis and Formey, a discussion over the legitimacy of different disciplines in making truth claims. This discussion shows that the discipline of metaphysics was, not only on a structural level but also on an ideological level, far

32. See Formey, 'Les preuves de l'existence de Dieu, ramenées aux notions communes', in *Histoire de l'Académie royale des sciences et belles lettres* (1747; Berlin, Haude and Spener, 1749), p.341-64 (348).

33. Formey, 'Les preuves de l'existence', p.343-44: 'Si les Mathematiciens ont passé pendant tant de siecles pour les seuls dépositaires de la certitude, c'est que leur Science etoit en effet la seule qui eut des principes incontestables, & qui ne les perdit jamais de vuë. [...] Ce n'est pas que les premieres notions de la Metaphysique soient moins claires, & gravées en caractéres moins ineffaçables dans nos esprits, que celles des Mathematiques. Mais elles sont plus difficiles à découvrir, parce qu'elles ont eu de tout tems & qu'elles ont encore aujourdhui à combattre, les plus redoutables ennemis de toute verité, les sens & l'imagination, qui offusquent ces verités dans la plupart des hommes; des Philosophes même, & qui arrêtent notre assentiment. Ajoutez qu'il ne decoule des verités Mathematiques aucune conséquence, qui interesse notre cœur & nos passions, au lieu qu'il n'en est pas de même des notions Metaphysiques, qui influënt nécessairement sur la Theologie Naturelle, & sur la Morale.'

from being the accepted foundation of all other disciplines practised at the Berlin Academy in the mid-1740s. Moreover, even Maupertuis's apparent general approval of the concept of metaphysics as the fundamental science could not conceal the increasing threat it faced from natural philosophy and mathematics.

Yet there was someone at the Academy who advocated much more explicitly than Maupertuis the superiority of natural philosophy and mathematical method and undermined the concept of metaphysics' role as fundamental science: Leonhard Euler, the Academy's renowned and prolific mathematician. As Leduc shows, Euler had the same idea of a metaphysics of nature or cosmology as the Academy's president, according to which cosmology was situated between *a priori* principles and experience-based laws of nature.[34] As in Maupertuis's case, this deprived metaphysical principles in general of their unconditional reliability. I would add to this that for Euler the purpose of this new conception of metaphysics seems to have been explicitly to discredit particular conceptions of metaphysics, mainly Wolff's. Through a much stronger emphasis on the important function of mechanics and mathematics in 'controlling' metaphysical *a priori* principles, in comparison with Maupertuis, Euler eventually paved the way for the dominance of natural philosophical and mathematical knowledge over a pure metaphysics.

In the same period when Maupertuis and Formey argued over the reliability of metaphysical methods in proving the existence of God, Euler was preparing another manoeuvre against the exclusiveness of traditional metaphysics. As he wrote in a letter to Maupertuis on 2 December 1747,[35] he had developed a proof to show that space and time were absolute and real; in his view, this was one of the most important questions of metaphysics. What is more, he boasted that his proof relied on mechanics, a discipline that only few of the metaphysicians could understand, which was why he expected fierce opposition from them if he should publish his work, particularly from those metaphysicians who adhered to the philosophy of Leibniz and Wolff. This apparently deliberately provocative treatise was presented by Euler to the Academy in February 1748 before it was published

34. See Leduc, 'Métaphysique de la nature', p.23.
35. See Euler to Maupertuis, 2 December 1747, in Leonhard Euler, *Commercium epistolicum*, vol.6: *Correspondance de Leonard Euler avec de Maupertuis et Frédéric II*, in *Opera omnia*, ed. Pierre Costabel, Eduard Winter, Ašot T. Grigorijan and Adolf P. Juškevič, series 4.a (Basel, 1986), p.95.

in the yearbook under the title 'Réflexions sur l'espace et le tems' ('Reflections on space and time'). The very first sentence of Euler's *mémoire* was a denial of metaphysics' quality as mother of all sciences: he declared that the principles of mechanics could not be inferred from general metaphysical rules; instead they gained their certainty and truth from their inherent coherence based on calculation.[36] This was an inversion of what Formey believed, based on Wolff's ontology, namely that mechanical laws of motion, just like mathematical axioms, could be deduced from the first notions of metaphysics. In contrast to Formey, Euler maintained that the conclusions attained in mechanics and natural philosophy in general should serve to adjust the first notions of metaphysics and hence help metaphysics in solving its questions:[37]

> since it is metaphysics that takes care of finding out the nature and properties of bodies, the knowledge of these truths could serve as a guide in this thorny research. Because one would have the right to reject in this science [metaphysics] all reasoning and all the ideas, which – no matter how well-founded they may appear – lead to conclusions that are contrary to these truths [the nature and properties of bodies]; and one would be authorized to accept only such principles that could persist together with these same truths. The first ideas that we have of the things external to us are usually so obscure and so sparsely determined that it is extremely dangerous to deduce certain results from them. It is thus always a great advantage when one knows already from elsewhere some conclusions to which the first principles of metaphysics shall lead: and it will be on the basis of these conclusions that one will have to adjust and to determine the first ideas of metaphysics.[38]

36. See Leonhard Euler, 'Réflexions sur l'espace et le tems', in *Histoire de l'Académie royale des sciences et belles lettres* (1748; Berlin, Haude and Spener, 1750), p.324-33 (324).
37. According to Euler's example, the real or concrete nature of space and time, as mechanics perceived it, contradicted the metaphysical notion of their mere imaginary character.
38. See Euler, 'Réflexions sur l'espace', p.324-25: 'comme c'est la Metaphysique, qui s'occupe à rechercher la nature & les propriétés des corps, la connoissance de ces verités pourra servir de guide dans ces recherches épineuses. Car on sera en droit de rejetter dans cette science tous les raisonnemens & toutes les idées, quelques fondées qu'elles puissent paroitre d'ailleurs, qui conduisent à des conclusions contraires à ces verités; & on sera autorisé de n'y admettre, que de tels principes, qui pourront subsister avec ces mêmes verités. Les premieres idées, que nous nous formons des choses, qui se trouvent hors de nous,

In this statement Euler alleged that the reason why metaphysical notions often had to be corrected by the results of natural sciences was that these notions were simply too obscure and indeterminate because of their *a priori* character. Hence, he used an objection against metaphysics that, to a lesser extent, had already been present in Maupertuis, and that denied the claim of metaphysics to be a source of certainty and clear evidence. This created a vacuum which mathematics as a certain science could fill.

Although, as Leduc has shown, Maupertuis and Euler still held on to the utility and epistemological legitimacy of metaphysics in relation to the natural sciences,[39] their scepticism towards the epistemological foundation of its first principles very much harmed the image of metaphysics' foundational function and of what had been declared to be the extraordinary characteristic of the Berlin Academy in 1745.

The adjustment of the hierarchy of disciplines

The detrimental consequences for the role of metaphysics as a discipline at the Academy, which Maupertuis's and Euler's criticism of metaphysical principles of the mid- and late 1740s had prompted, can be observed in the speech that Maupertuis gave in his function as president at the Academy's public assembly of 18 June 1750. Under the title 'Des devoirs de l'académicien' ('On the duties of the academician'), this speech allowed Maupertuis to redefine the hierarchy of the institution's four disciplines or classes.[40] This hierarchy formally arranged the four classes according to their subjects: first came the experimental philosophy class, whose members observed the sensible phenomena of the universe; it was followed by mathematics,

---

sont ordinairement si obscures & si peu déterminées, qu'il est extrémement dangereux d'en tirer des conséquences, dont on puisse être assuré. C'est donc toujours une grande avance, quand on connoit déjà d'ailleurs quelques conclusions, auxquelles les premiers principes de la Metaphysique doivent aboutir: & ce sera sur ces conclusions, qu'il faudra régler & déterminer les premieres idées de la Metaphysique.'

39. See Leduc, 'Métaphysique de la nature', p.28-29.

40. See Maupertuis, 'Des devoirs de l'académicien', in *Histoire de l'Académie royale des sciences et belles lettres, depuis son origine jusqu'à présent, avec les pieces originales* (Berlin, Haude and Spener, 1750), p.115-25. The text was reprinted in the 'Mémoires' section of the *Histoire* of 1753 (Berlin, Haude and Spener, 1755), p.511-21. For a translation of the text see the appendix to this volume. All direct quotations of 'Des devoirs de l'académicien' in this chapter are taken from this translation.

which examined the majority of objects that lay beyond the sensible phenomena of bodies; finally there was the class of speculative philosophy, which dealt with the objects that were entirely separated from physical bodies, that is with spirits. The belles-lettres class stood apart from these three, dealing with literature, language and history. At this level, Maupertuis's narrative of the hierarchy of the classes seemed not to be evaluative; he listed rather than assessed the different classes according to their thematic focus.

However, beyond this list, within the more detailed description of each class, the academicians were surely able to perceive how Maupertuis ranked the four classes according to questions of utility and evidence. In this ranking, geometry, that is mathematics, seemed to hold the dominant position for it had gradually extended its field of application and was now able to explain all phenomena of nature that beforehand escaped human understanding:

> Geometry, which owes its origin to its utility, and which the first geometers applied with so much success to the needs of life, was then for several centuries only sterile speculation and a kind of mind game. [...] The happy revolution brought about recently in our day in the sciences has rendered it more audacious. We have seen geometry explain all the phenomena of motion, and which part of this does not pertain to natural philosophy? We have seen it follow the ray of light in the space of the heavens, through all the bodies that it penetrates, calculate all the marvels which arise from its reflections and refractions; either to lead us to discover objects which because of their immense distance evaded our eyes, or to reveal to us those which by their extreme minuteness could not be perceived. [...] We have seen the mathematical sciences applied to all the useful or agreeable arts.[41]

41. See Maupertuis, 'Des devoirs de l'académicien', p.119: 'La Geometrie, qui doit son origine à son utilité, & que les premiers Geometres appliquerent avec tant de succés aux besoins de la vie, ne fut ensuite pendant plusieurs siecles qu'une spéculation sterile, & une espece de jeu d'esprit. [...] L'heureuse révolution qui s'est faite presque de nos jours dans les Sciences, la rendit plus audacieuse. On vit la Geometrie expliquer tous les phenomenes du Mouvement, & quelle partie n'est-ce pas de la Philosophie naturelle? On la vit suivre le Rayon de la Lumiere dans l'espace des Cieux, à travers tous les corps qu'il pénétre, calculer toutes les Merveilles qui naissent de ses réfléxions & de ses réfractions: soit pour nous faire découvrir des objets que leur immense éloignement déroboit à nos yeux, soit pour nous rendre sensibles ceux qui par leur extreme petitesse ne pouvoient être apperceus. [...] On vit les sciences Mathematiques s'appliquer à tous les Arts utiles ou agréables.'

According to this description, mathematics had achieved the transition from an isolated, completely abstract and theoretical science to a science of practical impact.

The same could not be said about metaphysics. In contrast to natural philosophy, to which mathematics could be applied so easily in Maupertuis's view, metaphysics dealt with such subjects as the supreme being, the human soul and the nature of spirits – subjects that precluded one from gaining certain knowledge of them by their very nature.[42] Maupertuis stressed that to apply mathematical method or concepts to such subjects, as some metaphysicians pretended to do, would not lead to the same degree of certainty as it did for the simple subjects that mathematicians dealt with[43] – a claim we have already noticed similarly in his 'Loix du mouvement'. As a consequence, Maupertuis advised the Academy's members to always subject their metaphysical reasoning to the *esprit philosophique*, an epistemological scrutiny that allowed them to distinguish whether the theories they established were evident, probable or dubious:

> If I exhibited here the severity of the danger of those speculations that concern the Supreme Being, first causes and the nature of minds [*esprits*], it is not, Gentlemen, because I wish to divert you from these researches. All is permitted to the philosopher, provided that he treats all with the *esprit philosophique*, that is to say, with that mind which measures the different degrees of assent, which distinguishes evidence, probability, doubt, and which only presents its speculations under that of these different aspects to which they belong.[44]

This epistemological plurality that Maupertuis seemed to attribute to metaphysical knowledge, however, precluded the possibility of metaphysics achieving a similar degree of practical impact and utility to mathematics, since the utility of mathematics was bound to the certainty of its theorems.

42. See Maupertuis, 'Des devoirs de l'académicien', p.120.
43. See Maupertuis, 'Des devoirs de l'académicien', p.120-21.
44. See Maupertuis, 'Des devoirs de l'académicien', p.121: 'Si je vous expose ici toute la grandeur du péril des spéculations qui concernent l'Etre suprème, les premieres Causes, & la Nature des Esprits, ce n'est pas M. M. que je veuille vous détourner de ces Recherches. Tout est permis au Philosophe, pourvû qu'il traite tout avec l'Esprit philosophique, c'est-à-dire, avec cet Esprit qui mesure les differens degrés d'Assentiment: qui distingue l'Evidence, la probabilité, le doute: & qui ne donne ses spéculations que sous celui de ces differens aspects qui leur appartient.'

Despite Maupertuis's (already previously expressed) scepticism concerning metaphysics' epistemological standing, he did not deny metaphysics' right to exist at the Academy. On the contrary, he used the extraordinary symbolic occasion of his speech – the Academy's public assemblies took place only twice a year and usually gathered not only ordinary and honorary members but also guests – to highlight once again the uniqueness of the institutions' speculative philosophy class, just as had been done in the Academy's first yearbook four years previously. Yet, in 1750, Maupertuis linked this uniqueness to the subjects with which metaphysics dealt: spirits and first causes. He stressed that, unlike in the past or in other academies, Frederick, the Academy's protector, did not ban research on these subjects: 'The legislators of all the academies, while ceding to them the entire nature of bodies, have forbidden that of spirits and speculation about first causes. Yet a monarch who has deigned to dictate our laws, a mind more vast, maybe more sure of your prudence, did not want to prohibit you from anything.'[45] By this, Maupertuis ascribed a rather important function to metaphysics: its presence was the reason why the Berlin Academy could be considered to be the freest academy in Europe.

However, this commendable function, since it was defined by subjects rather than by method, promoted a concept of metaphysics that was very different not only from the one displayed by Formey in the 1745 'Préface' and in his two treatises of 1747 on God's existence but also from Maupertuis's own treatise of 1746 on the Principle of least action. In his speech of 1750, Maupertuis no longer mentioned metaphysics' role as a science of universal principles from which the principles and eventually the observable phenomena of every other science could be deduced. On the contrary, it seems rather that Maupertuis had eventually conferred this role to mathematics, due to its impact on natural philosophy as he emphasised it. As a result of Maupertuis's speech on the duties of the academician, metaphysics' role as the mother of all sciences thus seems to have been reversed only a few years after its solemn proclamation.

45. Maupertuis, 'Devoirs de l'académicien', p.117: 'Les Legislateurs de toutes les Académies, en leur livrant la Nature entiere des Corps, leur ont interdit celle des Esprits, & la speculation des premières causes: Un Monarque qui a daigné dicter nos loix; un Esprit plus vaste, plus seur peut être aussi de vôtre prudence, n'a rien voulu vous interdire.'

The reinvention of metaphysics' legitimacy after mid-century

The shift of focus from the method of metaphysics to its subjects and the increasing distrust in metaphysics' ability to provide certainty were reflected in the works of the philosophers who joined the Berlin Academy around mid-century. These new members, mainly of Swiss provenance, still witnessed Maupertuis's influence as the Academy's president and hence, in their writings, they partially tried to explain and to refine his conception of metaphysics. At the same time, however, they were part of a new era at the Academy, in which Maupertuis's influence steadily declined due to his increasing absence from Berlin: after 1753 he only sporadically attended the Academy's assemblies because of health issues, and in 1759 he eventually died. Under these conditions, the new members seem to have abandoned the concern with empirical certainty that we have observed in Maupertuis's considerations on different disciplines and methods, and instead made an effort to preserve metaphysics' legitimacy as a science despite its epistemological and methodological limitations.

This can be seen firstly in a discourse entitled 'Les premiers principes de la métaphysique' ('The first principles of metaphysics') by the Swiss philosopher and professor of mathematics Nicolas Béguelin. He had been affiliated with the Academy since November 1747 and hence experienced, at least in part, the debates concerning the certainty of metaphysical knowledge inside the Academy. Béguelin, like Formey, had been trained in Leibnizian–Wolffian philosophy, but his own interests as a philosopher were multifarious: although he was a member of the speculative philosophy class, his *mémoires* spanned a much broader range of subjects, like mechanics, optics and meteorology. His first *mémoire*, for example, dealt with the hatching of eggs. Subsequently, he dedicated more than ten years and five academic treatises to the investigation into the nature and elements of metaphysics. As François Duchesneau has shown, at the core of these essays was his attempt to root experimental philosophy in basic metaphysical principles.[46]

Béguelin's treatise of 1755 was the start of this series of essays, and it was an attempt to explain why the discipline of metaphysics

46. See François Duchesneau, 'Nicolas de Béguelin et les fondements d'une philosophie de la nature', *Philosophiques* 42:1 (2015), p.89-105.

was often and maybe also justly discredited by his contemporaries. He described metaphysics as the only science which made very little progress and whose practitioners constantly engaged in controversy. According to him, this unfortunate situation had arisen because the first principles on which metaphysics based itself were not clear and therefore not universally accepted. This, in turn, made it difficult for metaphysics to uphold its acclaimed role as the foundation of all other sciences:

> A science that serves as the basis for all the others, that offers them the first principles, would it not have such [first principles] itself? Or would they [the principles] be so much out of the human mind's range that it might not be able to understand it; or if it catches them would it not be through clouds that are so dark that it [the mind] could never be sure to not be mistaken?[47]

The example he used to prove his point was the Leibnizian Principle of sufficient reason, and he demonstrated in much detail that this principle could not count as an axiom and hence did not possess the same sort of evidence as did mathematical principles. Nonetheless, Béguelin held the opinion that the Principle of sufficient reason was a crucial element within metaphysics that served as a basis for the most important truths:

> After everything I have said, one could be tempted to suspect that my aim is to weaken the truth of the Principle of sufficient reason: this is not at all my intention; I am strongly convinced that nothing is as it is without sufficient reason; I know and I have just shown that one would not be able to make a certain step – neither in philosophy, nor in metaphysics – if one does not use this principle as a basis for our reasoning; the most important and honourable truths depend on it, although it does not suffice for revealing them to us.[48]

47. Nicolas Béguelin, 'Les premiers principes de la métaphysique', in *Histoire de l'Académie royale des sciences et belles lettres* (1755; Berlin, Haude and Spener, 1757), p.405-23 (405): 'Une Science qui sert de base à toutes les autres, qui leur prête les premiers principes, n'en auroit elle point elle-même? Ou ceux-ci seroient-ils tellement hors de la portée de l'esprit humain qu'il ne pût les saisir; ou s'il les attrape, est-ce à travers des nuages si sombres qu'il ne puisse jamais s'assurer de ne s'y être point trompé?'
48. See Béguelin, 'Premiers principes', p.422: 'Après tout ce que je viens de dire, on seroit tenté de soupçonner, que mon but est d'ébranler la vérité du Principe de la raison suffisante: ce n'est nullement mon intention; je suis intimément persuadé que rien n'est comme il est sans raison suffisante; je sai, & je viens de

Béguelin's judgement on the Principle of sufficient reason operated with the same emphasis on subjects as that contained in Maupertuis's 'On the duties of the academician' for metaphysics in general. According to Béguelin, the truths or matters that this basic metaphysical principle tried to explain were so sublime that its explanatory limitations could not harm the legitimacy of the principle. Béguelin stressed that, in order to compensate for these methodological limitations and to avoid them discrediting the entire discipline, its practitioners should generally accept the limited degree of evidence that metaphysical principles bore, instead of overstretching the potential certainty of one's own preferred principle.[49] We can thus say that Béguelin's treatise of 1755 not only explained more precisely than Maupertuis before him what the problem with the perception of metaphysics consisted in, but it also introduced the first step towards a change of paradigm in this perception in the second half of the eighteenth century, away from the negative connotation of its limited degree of certainty towards its acceptance and eventual reappraisal.

This change was finally accomplished ten years later by his academic colleague Jean Bernard Merian who had joined the Academy in April 1750, where he seems to have quickly gained a reputation and was particularly close to the president, Maupertuis. His interest lay predominantly in British theories of cognition, but his academic treatises, like those of Béguelin, dealt with many diverse subjects and reveal his rather eclectic conception of philosophy.[50] In 1765, Merian was given the opportunity to make a speech at the Academy's public assembly on 6 June and he chose, like Maupertuis's 'On the duties of the academician', to deal with the relationship between the different sciences at the Academy. Yet, unlike Maupertuis's speech, Merian's 'Discours sur la métaphysique' ('Discourse on metaphysics') – as its title suggests – focused explicitly on metaphysics and achieved a re-establishment of metaphysics' outstanding role among the contemporary scientific

le montrer, qu'on ne sauroit faire un pas asseuré, ni dans la Philosophie, ni dans la Métaphysique, à moins de poser ce Principe pour base de nos raisonnemens; les vérités les plus importantes & les plus respectables en dépendent, quoiqu'il ne suffise pas à nous les découvrir.'

49. See Béguelin, 'Premiers principes', p.423.

50. See Christian Bartholmèss, *Histoire philosophique de l'Académie de Prusse depuis Leibniz jusqu'à Schelling, particul. sous Frédéric-le-Grand*, 2 vols (Paris, 1850), vol.2, p.32-76, and Jens Häseler, 'Johann Bernhard Merian – ein Schweizer an der Berliner Akademie', in *Schweizer im Berlin des 18. Jahrhunderts*, ed. Martin Fontius (Berlin, 1996), p.217-28.

disciplines in general and at the Berlin Academy in particular. After twenty years and after Maupertuis had been dead for six years, Merian finally explicitly reminded the academic public of metaphysics' original meaning as the first science which contained the elements of all other sciences.[51] He not only restated Béguelin's appraisal of the utmost relevance and sublimity of the objects studied in metaphysics, thus inferring the excellence of the discipline,[52] he also stressed, apparently in the footsteps of Formey, the methodological strength of metaphysics and its power to provide first notions to other sciences.[53] Nevertheless, Merian also made it clear that he did not want to rank the different disciplines but rather granted them all equal legitimacy. In his view, the sources of legitimacy were different for each discipline: one grounded its legitimacy in its outstanding objectives, the other in its reliable method:

> It seems to me that the value of each science consists in the goal for which it aims and of its capacity to reach it. Yet, if I am not mistaken, since these two aspects counterbalance each other, the comparative value of all human sciences will be reduced to a sort of parity. The one works on the more important subjects, the other with more ability: the first is more sublime for its opinions, the second is more certain for its operations.[54]

In order to illustrate his theory of compensation among the sciences, Merian compared experimental and speculative philosophy with each other. On the one hand, metaphysicians generated knowledge that went far beyond the rather superficial knowledge that resulted from physicists' observations. Furthermore, in contrast to what Euler had claimed some twenty years earlier, Merian argued that there existed findings in natural philosophy which had to be corrected by such

51. See Jean Bernard Merian, 'Discours sur la métaphysique', in *Histoire de l'Académie royale des sciences et belles lettres* (1765; Berlin, Haude and Spener, 1767), p.450-74 (451).
52. See Merian, 'Discours sur la métaphysique', p.452.
53. See Merian, 'Discours sur la métaphysique', p.453.
54. Merian, 'Discours sur la métaphysique', p.454: 'Le prix de chaque Science me paroît être en raison composée du but où elle tend, & de la faculté d'y atteindre. Or, si je ne me trompe, ces deux points étant compensés l'un par l'autre, la valeur comparative de toutes les sciences humaines se réduira à une espece d'égalité. L'une travaille sur des sujets plus importans, l'autre avec plus de facilité: celle-ci est plus sublime dans ses vues, celle-là plus sûre dans ses opérations.'

metaphysical knowledge (*spéculation*).[55] On the other hand, in terms of method, Merian had to admit that natural philosophy clearly was in a more advantageous position than metaphysics, since its objects of study – material bodies – were much better suited to observation and experiment than the complicated and non-tangible objects of metaphysical investigation. Metaphysics, therefore, really had no other choice than to rely largely on theoretical contemplation, which unfortunately often was not successful.[56]

In the same vein as comparing natural philosophy and metaphysics, Merian engaged also, like everybody before him, in a discussion of mathematics' role in all this. However, contrary to the unanimous positive appraisal of mathematics, Merian considered it as a mere auxiliary science that was nothing more than an ontology of quantities. Although originally dependent on metaphysics, mathematics had developed over time into a shining star in the house of science and had become the most important rival to metaphysics. This rivalry, as we have seen in all the academic discourses quoted earlier, had led to continuous comparisons regarding the degrees of certainty that these two sciences could provide, and especially to attempts to judge metaphysics according to its ability to provide mathematical certainty. Merian eventually broke with this tradition:

> This application of the mathematical method has been endlessly detrimental to the speculative sciences: it has often spread absurdity over the most respectable truths. [...] We would do an essential favour to reason if we could cure philosophers of this sickness by persuading them that there lies infinitely more science in knowing how to ignore what we don't know, than in knowing how to demonstrate it.[57]

With this condemnation of a 'mathematisation' of metaphysics he continued the efforts of Béguelin to release metaphysics from its dependence on mathematical evidence.

Regarded superficially, Merian's statement can easily be interpreted as simply another example in the two-sided controversy about the

55. See Merian, 'Discours sur la métaphysique', p.455.
56. See Merian, 'Discours sur la métaphysique', p.457.
57. Merian, 'Discours sur la métaphysique', p.460: 'Cette affectation de la méthode géométrique a fait un tort infini aux sciences spéculatives: elle a souvent répandu du ridicule sur les vérités les plus respectables. [...] On rendroit un service essentiel à la Raison, si l'on pouvoit guérir les philosophes de cette maladie, en leur persuadant qu'il y a infinement plus de science à savoir ignorer ce qu'on ignore, qu'à savoir le démontrer.'

reliability and certainty of metaphysics, more precisely about its demonstrative method. Yet I shall argue that he abandoned this controversy in order to emphasise a different quality of metaphysics. Hence, Merian's statement concerning the futility of applying mathematics to speculative sciences was not simply a one-sided attack on Wolffian metaphysicians who had been convinced of the absolute evidence that resulted from metaphysical demonstration and, because of this, were (mistakenly) considered as applying mathematics to metaphysics. On the contrary, I would argue that Merian's judgement also included a criticism of those philosophers who compared mathematics and metaphysics and came to a negative result, like Maupertuis. In my view, Merian wanted to eliminate every kind of association between mathematics and metaphysics as had been common at the Academy before mid-century. Instead, he suggested a new, completely opposite quality that was able to affirm metaphysics' legitimacy and strength: the quality to accept (and perhaps also to show) the moments of uncertainty or probability instead of desperately trying to render certain what was not. Moreover, he stressed metaphysics' merit of having shown us the limits of the human mind.[58]

In the same vein, Merian highlighted the crucial function of the diversity of opinions that had been developed and discussed in metaphysics over time, or its lack of one universal doctrine. It seems that it was this 'fermentation des esprits', as Merian called it, which sparked in his eyes the progress of metaphysical knowledge, not the demonstrative method. Probably he had precisely this quality of metaphysics in mind – to establish and ponder various different truth claims – when he eventually praised the *esprit philosophique* which metaphysics, according to him, was able to spread to all other sciences.[59] This *esprit philosophique*, or eclecticism, was the quality that bestowed again on metaphysics the role of a fundamental science since, according to Merian, it was through this *esprit philosophique* that metaphysics 'fertilised' and enriched the other sciences.

This significant shift in paradigm from certainty towards probability and eclecticism in the characterisation of the metaphysical method was surely a direct reaction to the outcome of the very popular prize contest that the Berlin Academy had set up only a few years before Merian's academic speech. In 1761, the Berlin academicians had asked

58. See Merian, 'Discours sur la métaphysique', p.464.
59. See Merian, 'Discours sur la métaphysique', p.464.

the public whether metaphysical truths were capable of the same sort of evidence as emerged from mathematics, and, if not, how high the degree of certainty of metaphysical truths was.[60] In 1763, the winner of the contest, the Jewish author Moses Mendelssohn, defended the similarity in certainty between mathematics and metaphysics, a position that – as we have seen – had been largely superseded within the Academy at least since Béguelin's treatise of the 1750s. By contrast, the runner-up (*accessit*), Immanuel Kant, argued for the separation of mathematics and metaphysics, and emphasised the genuine character of the philosophical method. The two winning essays seem to match the positions of the two different camps who had debated the role of metaphysics in the previous years: Mendelssohn's answer followed Formey's emphasis on the certainty of the metaphysical method, whereas Kant reflected Maupertuis's, Euler's and also Béguelin's theories of the difference between mathematical evidence and metaphysical demonstration.[61]

As we have seen, in his speech, Merian overcame the controversy that had animated the Academy during the previous years, and hence aligned himself with neither of the two answers to the essay contest. Instead, Merian's speech seems to contain a synthesis of several different arguments that the contest had sparked, particularly from the non-distinguished anonymous proposals. As Paola Basso has highlighted, many of these answers had challenged the deeply rooted assumption both of mathematics' absolute certainty and of the superiority of the demonstrative method in metaphysics.[62] For example, Merian claimed that, similar to metaphysics, natural philosophy and mathematics were not free of uncertainty. More precisely, he criticised practitioners of these two disciplines for taking for granted certain principles that, once they were pared down to

60. The question was declared on 4 May 1761 in the Academy's assembly: 'On demande si les vérités de la Métaphysique en général, et en particulier les premiers principes de la Théologie naturelle et de la Morale, sont susceptibles de la même évidence que les vérités mathématiques, et au cas qu'elles n'en seraient pas susceptibles, quelle est la nature de leur certitude, à quel degré elles peuvent parvenir, et si ce degré suffit à la conviction.' The winner of the essay contest was chosen on 26 May 1763.

61. See Prunea-Bretonnet, 'La méthode philosophique en question', p.107-30.

62. See Paola Basso, 'Deutlichkeit, évidence et certitude dans les réponses anonymes', *Astérion* 9 (2011), doi:10.4000/asterion.2101, and 'Le mythe de la démonstrabilité résiste-t-il encore? Remarques sur l'orientation des réponses anonymes', *Astérion* 9 (2011), doi:10.4000/asterion.2112.

their essence, lost all their clarity.[63] With regard to metaphysics, he claimed that it was wrong to assume that metaphysical objects were entirely provable and to regard the method of demonstration as the only means of providing certainty.[64] These two aspects posed a crucial challenge to the fundamental and widely acknowledged assumptions on which the entire controversy about the concept of metaphysics relied.

Moreover, the particularity of Merian's text, compared to the discussion that evolved within the prize contest, was that he not only reflected the modified concepts of clarity and certainty that began to mark metaphysics and its neighbouring disciplines in the 1760s, but he also translated them into considerations concerning the (future) role of metaphysics as a discipline at the Berlin Academy and beyond. In Merian's view, in his day the conditions for the discipline of metaphysics to flourish were ideal: first, because theology had become more tolerant towards metaphysical interpretations of doctrine, and second, because learned societies were allowed to freely discuss metaphysical topics.[65] Consequently, Merian conferred the whole responsibility for benefitting from this situation onto the practitioners of metaphysics and exhorted them to lead the discipline to the excellence it was able to reach. The metaphysicians, however, had to change their practice to allow for this excellence: they had to give up their adherence to particular systems and schools of metaphysics and their immoral custom of engaging in strife over theories:

> This zeal for opinions and authorities, which the education in systems inspires, fills the mind with venom and acidity and lights the fire of dissension. How could a man, who during his whole life did nothing except demonstrating and who never learned to doubt, accept that he is refuted or that others hesitate to adopt his conclusions? No matter which system you embrace, you will have all those who follow different banners against you; and if you do not embrace any [system] the united forces of all systems will swoop on you.[66]

63. See Merian, 'Discours sur la métaphysique', p.461.
64. See Merian, 'Discours sur la métaphysique', p.462.
65. See Merian, 'Discours sur la métaphysique', p.465.
66. See Merian, 'Discours sur la métaphysique', p.467: 'Ce zele pour les opinions, & pour les autorités, que l'éducation systématique inspire, remplit, pour l'ordinaire, l'esprit de fiel, & d'aigreur, & allume le tison de la Discorde. Comment un homme qui toute sa vie n'a fait que démontrer, & qui n'a jamais sû douter, souffrira-t-il qu'on le contredise, ou qu'on balance à adopter ses assertions? Quelque systeme que vous embrassiez, vous avez pour ennemis tous

Merian's statement was certainly an echo of the anti-system debate which had peaked in France in mid-century with the appearance of Condillac's *Traité des systèmes* (*Treatise on systems*) (1749) and D'Alembert's considerations in the 'Discours préliminaire' ('Preliminary discourse') in the *Encyclopédie* (1751).[67] The hypothetical and overly abstract nature of metaphysical systems, which was at the centre of those authors' criticism, had also been the stumbling block in the discussions about metaphysics at the Berlin Academy that I have outlined throughout this chapter. Thus, in 1765 Merian also most likely had the developments in mind that had occupied his own institution over the past twenty years when he warned his colleagues about the dangers and folly of adhering to one particular system. Instead of vindicating one school of thought, the good philosopher, and the academician in particular, should rather seek an eclectic metaphysics that draws on sources from various different fields and focuses on contemplation.[68] From Merian's speech we might conclude that his emphasis on metaphysics' speculative (in the sense of hypothetical) character not only re-established the recognition of metaphysics as a universal science, but also calmed the debate and enabled free and non-doctrinal scientific practice, appropriate for a learned society.

Epilogue: back to the start?

By way of conclusion we might ask how far Merian's vision for an eclectic and genuinely hypothetical metaphysics reflected the reality at the Berlin Academy twenty years after its renewal. Did his colleagues approve or even practise the guidelines that Merian presented in 1765? In this respect it is particularly curious that the same yearbook in which Merian's public speech on metaphysics was published also contained an intervention by Formey that mirrored some of Merian's considerations. Formey's treatise, similar to those he had presented in 1747 in the debate with Maupertuis, asked about the legitimacy of different *a posteriori* proofs for the existence of God,

ceux qui suivent d'autres étendarts; & si vous n'en embrassez aucun, les forces réunies de tous les systemes vont fondre sur vous.'

67. See Peter R. Anstey, 'The principled Enlightenment: Condillac, D'Alembert and principle minimalism', in *Rethinking the Enlightenment: between history, philosophy, and politics*, ed. Geoff Boucher and Henry Martyn Lloyd (Lanham, MD, 2018), p.131-50.

68. See Merian, 'Discours sur la métaphysique', p.469.

and through this subjected the disciplines of natural philosophy, mathematics and metaphysics to scrutiny. As a criterion for judging the appropriateness of each discipline he again posed its degree of certainty. Pure natural philosophy, relying only on the observation of nature, he found too superficial; mathematical physics was no better in his opinion, in the sense that it produced the same knowledge as natural philosophy but disguised it with calculation. Concerning mathematical metaphysics, however, Formey was more indulgent, for it relied on the method of demonstration which Formey still considered the most superior means to achieve truth: 'Nothing, hence, would be more fortunate and more desirable than a clear demonstration which is concise and shielded from all contestation, which would be able to establish the existence of God on unique and unshakeable grounds.'[69] It seems that, in contrast to Merian, Formey had still not given up the idea of a metaphysics that had unlimited access to truth in every aspect of the spiritual world.

In 1765, Formey's claims concerning the concept and role of metaphysics were the same as in the 1740s, although the context had changed. He still described metaphysics as the science that established general notions through logical reasoning and abstraction that eventually served as the basis and guiding principles for other sciences. Hence, in his view, all arguments for the existence of God, whether they were inspired by natural philosophy, mathematical natural philosophy or mathematical metaphysics, were only true as far as they were followed through to their metaphysical foundations.[70] Put like this, it was not difficult for Formey to claim that metaphysics had the exclusive right to decide on questions concerning divinity. Moreover, Formey's essay explicitly showed that, within the discipline of metaphysics, he considered only one particular approach valid, namely the Leibnizian–Wolffian argument in which God's existence was inferred from the contingency of the world. It seems, then, that Formey ignored Merian's appeal to the academicians to practise an

69. See Formey, 'Discours sur ces questions: Quel est le degré de certitude dont sont susceptibles les preuves tirées de la considération de cet univers pour demontrer l'existence d'une divinité? Et quelle est la meilleur maniere de faire usage de ces argumens a posteriori, pour établir cette importante vérité?', in *Histoire de l'Académie royale des sciences et belles lettres* (1765; Berlin, Haude and Spener, 1767), p.435-49 (441): 'Rien donc ne seroit plus heureux & plus desirable qu'une Démonstration nette, concise, & à l'abri de toute contestation, qui établiroit l'existence de Dieu sur un fondement unique & inébranlable.'

70. See Formey, 'Discours sur ces questions', p.443.

eclectic metaphysics that was open to several different approaches. It is even possible that Formey deliberately inserted his treatise into the same yearbook in which Merian's public speech was published, in order to demonstrate his opposition to his colleague's suggestions.[71] If in the 1740s Formey had used these arguments to fight against the marginalisation of metaphysics through the increasing emphasis on mathematics and natural philosophy, in 1765 he opposed what he might have considered an internal decomposition of the discipline. Hence, Merian's reformation of metaphysics – despite the confirmation of the outstanding role of metaphysics that it contained – did not find unanimous support among the members of the Berlin Academy. Twenty years after the unprecedented establishment of the class of speculative philosophy at the Berlin Academy, the justification of metaphysics' importance was still controversial.

71. However, Formey had written his treatise much earlier, namely, in late 1754 or 1755. It had been his contribution to the 1755 prize essay contest run by the Academy of Leiden and had been published for the first time in Formey's journal *Nouvelle bibliothèque germanique* 25 (1759), p.317-41.

Eclectic philosophy and 'academic spirit': the Berlin Academy and the Thomasian legacy

TINCA PRUNEA-BRETONNET

The late 1750s and early 1760s represent a turning point in the German Enlightenment. Several crucial factors shaped the philosophical life of the period, such as Christian Wolff's death in 1754, the relative decline of his school despite the academic success of some followers, and the lively debates triggered by controversial prize questions published by the Berlin Academy, to name but a few. A new balance emerged within the Berlin Academy itself in the aftermath of P.-L. Moreau de Maupertuis's death in 1759. A dynamic and prominent actor on the philosophical scene since its reorganisation by Frederick the Great in 1744/1746, the Academy had been characterised by the confrontation between two camps: on the one hand, a Newtonian and anti-Wolffian group led by the president, Maupertuis, promoting French and English enlightened ideas, and, on the other, a less powerful Wolffian faction led by J. H. S. Formey and J. G. Sulzer. Wilhelm Dilthey compared them to 'two rivers' that flow in the same bed for years but never merge.[1] Since the two orientations did influence each other from the beginning and key members occasionally sided with the opposite camp, this image is perhaps too strong, but it nevertheless captures the general direction of their collaboration.

The first claim of this chapter is that starting in the early 1760s a momentous change occurs within the Academy, a change that

My first interest in J. B. Merian was generously supported by the IZEA Halle (Saale) through a fellowship in 2015. I would like to warmly thank this institution for the fruitful time spent at the Centre. This chapter is published within a grant funded by the Ministry of Research, Innovation and Digitization, CNCS/CCCDI – UEFISCDI (project number PCE 105/2021, within PNCDI III).

1. Wilhelm Dilthey, 'Friedrich der Grosse und die deutsche Aufklärung', in *Studien zur Geschichte des deutschen Geistes*, *Gesammelte Schriften*, vol.3 (Leipzig and Berlin, 1927), p.81-157 (135).

attests to the elaboration of a specific academic philosophy. Dilthey identified what he took to be a newfound consonance, namely, a common interest in moral and political issues. While not denying this, I claim that the salient feature of this unity is the articulation within the Academy of a specific theoretical approach, which amounted to a novel 'art of philosophising'. I argue that by 1765, with the appearance of Jean Bernard Merian's *Discourse on metaphysics*, an academic philosophy, based on eclectic methodological principles – and indeed later claiming this very label – attains a certain maturity and coherence, notwithstanding divergences about particular points of doctrine. The debates about the fate of metaphysics occupy centre stage and help crystallise the main theses of this movement. I therefore see the academic, eclectic philosophy, unlike the so-called *Popularphilosophie* of the same period, as focused on theoretical issues, rather than moral questions. This theoretical emphasis allows us to distinguish between these two contemporary movements, which are too often lumped together, and to make the case for the autonomy of the perspective endorsed within the Academy.

The second claim of this chapter is that there was a strong continuity between the academic, eclectic philosophy and Christian Thomasius' views. I show how academic philosophers such as Merian and Formey took up several central theses advanced by Thomasius and, in so doing, not only grounded their approach in the *Frühaufklärung*, but, furthermore, made it their aim to continue the programme of the German Enlightenment by promoting one of its central ideals, namely, the imperative to think for oneself.[2] It is high time to revise the familiar narrative which has for too long relegated the Berlin Academy to the status of an 'intellectual Parisian colony on German ground',[3] of a foreign body within the *Aufklärung*.

The aim of this chapter is therefore twofold: first, to highlight the main features of the self-defined 'eclectic philosophy' of the Berlin Academy as an autonomous and philosophically robust perspective on metaphysics; and second, to demonstrate that this form of eclecticism belongs to the German Enlightenment as such, and is an integral part

2. See Immanuel Kant, 'What is orientation in thinking?' (1786), in *Kant's political writings*, ed. Hans Reiss (Cambridge, 1991), p.237-49 (249): '*Thinking for oneself* means seeking the supreme touchstone of truth in oneself (i.e. in one's own reason); and the maxim of always thinking for oneself is *enlightenment*.'
3. Max Wundt, *Die deutsche Schulphilosophie im Zeitalter der Aufklärung* (1945; Hildesheim, 1992), p.321.

of the philosophical movement initiated by representative thinkers such as Christian Thomasius. It therefore offers a revisionist reading of this period against the widespread tradition which overlooks the Berlin Academy as insignificant for the Enlightenment in general or even alien to German philosophy.

The first four sections present J. B. Merian's perspective on metaphysics and his eclectic methodology as articulated in the *Discourse on metaphysics*. I then analyse Formey's and Maupertuis's views on academic philosophy. The final section discusses how academic philosophy relates to Thomasius' philosophical programme and to the corresponding treatment of eclectic philosophy in J. G. Walch and J. J. Brucker. From this analysis I conclude that the eclectic philosophy promoted by the Berlin Academy is consistent with the tradition of the *Frühaufklärung* and, from the 1760s, represented an autonomous movement within the German Enlightenment.

The fate of metaphysics

Merian's *Discourse on metaphysics* was read at the Academy in 1765 and published as a stand-alone piece the same year, and again two years later in the *Histoire*.[4] As stated in the preface, it represents an early and shortened answer to the *Preisfrage* for 1763 on method in philosophy, written before Merian assessed the essays submitted by the participants in the academic contest. I argue that this text is, if not the first, then certainly one of the most representative essays of the academic philosophy elaborated by the Berlin Academy.

Merian had been a member of the class of speculative philosophy since 1750, and was close to president Maupertuis and to his philosophical convictions. Unlike several of his colleagues, he was fluent not only in French, the official language of the Academy, but also in German (his mother tongue), English and Italian. Proficient in Latin and Greek as well, he was an erudite and prolific writer and translator. Merian may be considered a member of the anti-Wolffian front led by Maupertuis; indeed, during several philosophical controversies he openly sided with the president. However, his position as a linguistic bridge within the Academy, as well as his bright and joyful character and vast knowledge, ensured that he played a somewhat mediating and multifaceted role among his colleagues. This is

4. Jean Bernard Merian, *Discours sur la métaphysique* (Berlin, Voss, 1765) (hereafter *Discourse*). All translations in this chapter are my own unless otherwise indicated.

manifest, for instance, in his genuine interest in, and admiration for, German thinkers like Lambert, Garve, Meiners and especially Mendelssohn.[5]

His *Discourse* is a tentative answer to the crisis of metaphysics, a common theme at that time with post-Wolffian thinkers. Merian attempts to elaborate a novel approach to metaphysics, capable of saving it from the disdain and perpetual conflicts that had prompted his contemporaries either to criticise and reject this discipline or, on the contrary, to lament its ruin. Two main goals stated in the preface guide his enterprise and announce his methodology: (a) to 'do justice' to all well-known philosophers irrespective of their nationality or philosophical school; and (b) to denounce any 'adulation' for living thinkers. This claim of impartiality and freedom of thinking and the rejection of authority are two central and distinct eclectic features which point to the perspective supported in this chapter.

The *Discourse*, Merian explains, attempts to assess the intrinsic and comparative value of metaphysics by analysing its 'advantages' and 'disadvantages', as well as the 'shortcomings' of those who practise it.[6] However, his explicit aim is to make the case in 'praise' (*éloge*) of metaphysics, for Merian is convinced that metaphysics 'is the most sublime and most excellent of all sciences'.[7] The first part of the essay establishes two indisputable theses on metaphysics, namely its primacy and its specificity, by means of a comparison with the experimental sciences and mathematics, while the second part discusses the main methodological elements of what Merian takes to be the 'true' art of philosophising.

Two standpoints are identified as responsible for the current deplorable state of metaphysics: first, the alleged 'enlightened' view of the 'ordinary person' (*le vulgaire*) who looks for material outcomes and visible results, and thus applies the pattern of practical sciences (such as optics, astronomy or mechanics) to metaphysics; second, the sceptical attitude which rejects any ambitious enterprise that challenges already acquired knowledge, our limited cognitive

5. See Merian's 'Abridgement of Moses Mendelssohn's prize-winning essay on the distinctness of the principles of metaphysics and natural theology read at the public meeting of the Prussian Royal Academy of Sciences on 2 June 1763 by M. Merian', in Immanuel Kant, *Theoretical philosophy, 1755-1770*, ed. and translated by David Walford and Ralf Meerbrote (Cambridge, 1992), p.276-86.
6. Merian, *Discourse*, p.5-6.
7. Merian, *Discourse*, p.6.

capacities, or simply everyday indolence. A third and powerful enemy of metaphysics comes from within: it consists, according to Merian, of the endless conflicts among metaphysicians.[8] The latter claim absolute knowledge for themselves, reject other systems, and strive to establish schools in order to exclusively promote their own doctrine. Consequently, any reform of or even reflection on the destiny of metaphysics would also have to address this latter crucial challenge, and Merian devotes the most important part of the *Discourse* to the analysis of the intrinsic perils and of the strategies to overcome them.

The primacy and specificity of metaphysics

On Merian's account, cognition has its origin in sense perception and observation;[9] individual material things are its first objects. He takes Newtonian physics as an established science, firmly founded on phenomena and as 'inexhaustible' as nature itself. On this basis, he presents the gradual elaboration and articulation of the sciences, starting with physics and continuing with cosmology. The latter examines the laws and concepts of physics (such as matter, motion, forces and simple beings) and deals with a 'transcendent world' where the first mechanisms of the universe are 'hidden' from our senses. This science is followed by psychology, which analyses human cognition and its faculties and reflects on the nature, origin and destination of humans. From psychology, through the reflection on the 'common origin' of bodies and souls, we move on to natural theology, which deals with the first cause or the eternal, infinite and independent being, and examines its attributes and relation to the world. Lastly, a crucial science comprising species and genera is formed by abstraction. It puts in order the ideas of the human mind and contains, according to Merian, the elements and the materials of all the sciences. Even if it is established after the other sciences,

8. 'If this science has not yet all the success it could have, if it is not respected as it should be, and if there still remain against it regrettable impressions – we should not blame it but rather those who practise it'; Merian, *Discourse*, p.35.
9. On Merian's views on empiricism and knowledge aquisition, see Tinca Prunea-Bretonnet, 'On the mitigated phenomenalism of J.-B. Merian', in *The Experiential turn in eighteenth-century German philosophy*, ed. Karin de Boer and Tinca Prunea-Bretonnet (New York, 2021), p.203-27, and Daniel Dumouchel, 'Jean Bernard Mérian: métaphysique et empirisme à l'Académie de Berlin', in *Académies scientifiques européennes et modernité*, ed. Pierre Girard, Christian Leduc and Mitia Rioux-Beaulne (Paris, forthcoming).

it plays a foundational and eminent role and this is why the name of 'first science', 'ontology' or 'metaphysics' in a narrow sense is still appropriate in his eyes.

This classification of the sciences, instituted by Wolff, was rather common at the time: metaphysics, in the broad sense, comprised ontology, cosmology, psychology and natural theology. In a narrow sense, general metaphysics was identified with ontology alone, followed by the other three disciplines, considered as domains of special metaphysics. Merian endorses here the Wolffian division and order of the philosophical sciences,[10] without however quoting his predecessor by name.

The combined criteria to assess the 'value' of each science are, according to Merian, the dignity of its goal and the probability of attaining it. If this double point of view is taken into account, he states, all the sciences are to be regarded as more or less equal. Metaphysics, for instance, has higher and more important goals, while the natural sciences advance in an 'easier' and more secure way. However, he believes that competition and rivalry among the sciences are useless and even harmful. Interconnected and virtually equal, the sciences and particularly the philosophical disciplines should collaborate and jointly ensure the progress of knowledge, which for Merian is yet to be attained.

However, this collaboration needs to be founded on two positive theses: the primacy and the specificity of metaphysics. For Merian, metaphysics ought to be acknowledged as the 'queen of the sciences' because it 'eminently contains all the sciences'.[11] Indeed, in the *Discourse* metaphysics is said to include not only the above-mentioned disciplines (ontology, cosmology, psychology, natural theology), but also mathematics, defined as the 'metaphysics of quantity', and to ground or support morals.[12] Aligning himself with an established perspective going back to Antiquity, Merian argues that metaphysics makes possible all the other sciences: 'From Aristotle to Leibniz all great men have admitted that its valuable influences flow into all the

10. 'We are following here the most received usage'; Merian, *Discourse*, p.5. On the order of metaphysical disciplines in Wolff, see Christian Wolff, *Discursus praeliminaris de philosophia in genere / Einleitende Abhandlung über Philosophie im Allgemeinen. Historisch-kritische Ausgabe*, ed. Günter Gawlick and Lothar Kreimendahl (Stuttgart and Bad Cannstatt, 1996), §99.

11. Merian, *Discourse*, p.28-29.

12. Merian, *Discourse*, p.20 and 6.

sciences, providing them with adequate principles, guiding notions, fair views and the correct method.'[13] Metaphysics not only explains the concepts used in the other sciences, but also 'rectifies' them, and indicates their limits; if we fail to appeal to this higher science, Merian believes, we inevitably fall into error. He ascribes to metaphysics alone the merit of bringing about an enlightened period and the duty to continue this work, rather than – as might have been expected – to the triumphant new physics or to geometry and its compelling ideal of certainty: 'If this century is less superstitious, more enlightened and more philosophical than the previous centuries, we principally owe it to it [i.e. to metaphysics].'[14] He unequivocally advocates its pre-eminence.

Notwithstanding this, Merian admits that the triumph of experimental sciences in the eighteenth century is indisputable. Moreover, the overpowering example of mathematics and the yearning to compete with it have cast disrespect on metaphysics: the former is clearer, more certain and evident, and its concepts are defined precisely and unambiguously. But, according to Merian, this comparative weakness of metaphysics should not be accepted, nor should it be mended by a harmful imitation of these sciences, because this would bespeak a misunderstanding of the very nature of this science. On the contrary, a sharp distinction should be drawn here: even if physics and metaphysics are both founded on observation and experience, their objects, method, relative certainty[15] and chances of progress are anything but similar. The same separation applies with regard to mathematics; the wish to imitate this science and to attain geometrical evidence has caused 'infinite damage' to metaphysics. The struggle or 'rage' to demonstrate becomes an 'illness' when imported into metaphysics, whose objects, Merian claims, are unfit for demonstration, thus endorsing an anti-Wolffian standpoint widely

13. Merian, *Discourse*, p.7: 'Depuis Aristote jusques à Leibnitz tous les grands hommes ont reconnu qu'elle verse ses influences heureuses dans toutes les sciences, qu'elle leur fournit des principes justes, des notions directrices, des vûes saines, & la vraie méthode.'
14. Merian, *Discourse*, p.33.
15. 'It's a mistake to think it [metaphysics] can be demonstrated; it's also a mistake to think it is entirely incompatible with geometrical evidence; it's an even bigger one to imagine that there is no certainty without demonstration' ('C'est un travers de la croire toute démontrable; c'en est un autre de la croire toute incompatible avec l'évidence géométrique; c'en est un plus grand de s'imaginer qu'il n'y a point de certitude sans demonstration'); Merian, *Discourse*, p.29.

shared at that time.[16] Metaphysics ought to be judged by different criteria; its specific type of certainty and evidence and its proper task and goals are no less honourable or convincing.[17] In short, Merian defends here a point of view typical of the opponents of Wolffianism, dominant at the Berlin Academy, while also shared by independent thinkers, such as Kant.[18]

Despite its deplorable current state, Merian is convinced that a new dawn has risen for metaphysics[19] and wishes to advance the means able to save it: a set of specific methodological principles which belong, as we shall see, to the eclectic art of philosophising.

A different approach to metaphysics: the 'philosophical spirit'

After having assessed the pre-eminence and the autonomy of metaphysics as the two positive foundations of his approach, Merian goes on to articulate a methodological programme centred on the development of what he calls the 'philosophical spirit' or the 'philosophical sense'. Merian considers the 'philosophical spirit' a 'gift of nature', given to virtually everyone, which needs to be nurtured, developed and 'turned into a habit'. Significantly, he argues that it should be sustained 'with first principles or [with] what is metaphysical in the sciences', highlighting once more the primacy of metaphysics among the sciences. It is described as a 'sentiment of truth' (*sentiment du vrai*) or a 'sagacity' which perceives what is true, chooses the shortest way, grasps the picture of nature as a whole, and attains straightaway the result that labour and diligence reach slowly and painfully:

> There is a sentiment of truth, a sagacity, which brings together distant connections [...]. This philosophical sense germinates and fertilises our ideas: it raises the natural philosopher above the mere observer, who, so to speak, organises the heavy mass of erudition, and it distinguishes the man of genius from the laborious scientist in all domains. This sentiment, I confess, is a precious gift of nature, which

16. Merian, *Discourse*, p.24.
17. See Tinca Prunea-Bretonnet, 'La méthode philosophique en question: l'Académie de Berlin et le concours pour l'année 1763', *Philosophiques* 42:1 (2015), p.107-30 (118-22).
18. See Kant's prize essay for the year 1763, 'Inquiry concerning the distinctness of the principles of natural theology and morality', in *Theoretical philosophy*, ed. and translated by D. Walford and R. Meerbrote, p.245-75.
19. Merian, *Discourse*, p.34.

> must nevertheless be cultivated; it is only by nourishing it with first principles or [with] what is metaphysical in the sciences that it can be developed and turned into a habit.[20]

Not only does the philosophical sense provide a comprehensive image of nature along with the possibility of isolating and analysing specific details, but it first and foremost enables the thinker to discern and choose the true and good elements from various doctrines. This is the reason why it can be regarded as fundamental for Merian's perspective which, as we shall see, promotes thinking for oneself, free but well-grounded choice, and the autonomous elaboration of doctrines – that is to say, it endorses the very principles of eclectic philosophy.

The development of the philosophical spirit requires an initial, negative step, which consists, according to Merian, in a threefold refusal: a refusal of philosophical systems, of sects and of blind reverence for authority. Two further methodological imperatives are closely related: the rejection of the ambition to demonstrate and to attain geometrical certainty in metaphysics, as well as the acceptance of our knowledge as partial and limited, of our condition as mere 'atoms in the immense city of the world, [where] our sphere is but a point'.[21]

On Merian's account, multifaceted and lasting damage has been caused by the urge to build systems. Their authors wrongly believe they possess absolute knowledge, and therefore fail to question their own ideas; after a lifetime of demonstration and system building, they end up being unable to tolerate contradiction or competition. Systems, he argues, prove even more harmful for disciples and epigones: taking for granted what they are taught, these young spirits and mediocre followers encourage the system builder to create a school or, more precisely, 'a sect' exclusively dedicated to the teaching of his doctrine. Philosophical sects, according to Merian, generate conflicts and even wars; on closer inspection, he claims, we see that in fact all religious

20. Merian, *Discourse*, p.7-8: 'Il est un sentiment du vrai, une sagacité, qui rapproche les rapports éloignés [...]. Ce sens philosophique fait germer & fructifier toutes nos idées: c'est lui qui éleve le physicien au-dessus du simple observateur, qui organise, pour ainsi dire, la masse pesante de l'érudition, & qui dans tous les genres distingue l'homme de génie du savant laborieux. Ce sentiment, je l'avoue, est un don précieux de la Nature, mais qui demande à être cultivé: & ce n'est qu'en le nourrissant des premiers principes, ou de ce qu'il y a de métaphysique dans les sciences, qu'on le développe, & on le fait tourner en habitude.'

21. Merian, *Discourse*, p.37.

and scientific wars spring from metaphysical debates. Members of sects necessarily lose their autonomy, their freedom of choice and finally their capacity to reason. They take a doctrine for the whole and only truth, despise other thinkers and cultivate an antagonistic attitude towards any other conception. Becoming 'gladiator metaphysicians', fanatics and enthusiasts, they set up never-ending disputes and deeply harm philosophy by bringing about what Merian calls, in Hobbesian terms, a generalised war in this science.[22]

All these deplorable effects are attributable to the fact that metaphysics is a particularly fertile field for human pride, self-love and self-sufficiency. It can bring out the best and the worst in a thinker. On Merian's account, 'the study of metaphysics, if misguided, may change into poison', because it is apt to combine error and vice.[23] The delusional and bellicose impression of owning the truth creates a 'scholarly prejudice' that is far more dangerous than the popular prejudice his age wishes to combat. His eclectic standpoint denouncing the adherence to sects as incompatible with the 'true' metaphysical attitude is accompanied here by the explicit rejection of systems. While shared by anti-Wolffian philosophers and by several French members of the Academy, this view is not necessarily found among the common eclectic theses of the period, with thinkers like Budde and Brucker, for instance, promoting the erection of systems. Merian, however, advocates their rejection because they reflect a pretentious desire to offer an all-embracing and exclusive view of the world, which is, in fact, inaccessible to our limited knowledge; he thereby echoes a Thomasian standpoint, as we shall see.

A positive approach should nevertheless accompany this negative programme: already existing systems and past doctrines are to be treated with respect and interest, Merian argues, and ought to be carefully studied. Even if systems have revealed their weaknesses, they may contain positive elements, and therefore one should take what is best from them. Certainly, this is to be done not by blindly and enthusiastically endorsing their ideas, as no authority or doctrine deserves adulation, but by becoming acquainted with them in a generous yet critical manner:

> There is not one of these famous men who has not triumphed over prejudices, established useful truths, opened new roads, or flattened

22. Merian, *Discourse*, p.40.
23. Merian, *Discourse*, p.43.

> the old ones. [...] I however acknowledge the respect owed to these creative geniuses who have advanced these large views and drawn these vast plans. Even if their systems may be only beautiful attempts, for this fact alone they are worthy of our admiration.[24]

This injunction is an important aspect of his philosophical standpoint. Merian, himself a learned and erudite philosopher, shows admiration for and interest in other authors, but always in a free and critical way. He will indeed stay true to this approach throughout his life. On his account, it is crucial to position one's thought within a venerable lineage of authors, from Plato and Aristotle up to Locke, Newton, Leibniz and Maupertuis. Irrespective of their nationality or philosophical sensibility, Merian considers them our 'forerunners and our models'; it is only by relying on their writings that we may cultivate the true philosophical spirit.[25] We may note that he also mentions here Descartes, Leibniz and Wolff, who were not held in particularly high regard within the Maupertuisian camp at the Berlin Academy.

All these past figures have contributed in their specific ways to this science, at least by showing the limits of our cognitive faculties, if not by shedding more light on the obscure regions of metaphysics. Consequently, on his account, the history of philosophy serves as an indispensable propaedeutic to any philosophical enterprise, and henceforth belongs to philosophy itself:[26]

> It is a vast gallery [of philosophical figures], expanding from east to west and from south to north. Sometimes our spirit enjoys following the order of its pictures, sometimes it likes to get lost in their variety. And what could possibly be more interesting to it? [This gallery] is its own history; it is the accurate mirror where the spirit can contemplate its smallness and grandeur, its rise and falls.[27]

24. Merian, *Discourse*, p.31 and 38: 'Il n'est aucun de ces hommes célebres qui n'ait vaincu des préjugés, établi des vérités utiles, ouvert de nouvelles routes, ou applani les anciennes. [...] Je connais cependant le respect dû aux genies créateurs qui ont proposé ces grandes vûes, & compassé ces vastes plans. Quand leurs systêmes ne seroient que de belles tentatives, ils mériteroient, à ce seul titre, toute notre admiration.'
25. Merian, *Discourse*, p.29.
26. On the relationship between history and eclectic philosophy, see Helmut Holzhey, 'Philosophie als Eklektik', *Studia Leibnitiana* 15:1 (1983), p.19-29 (25-28).
27. Merian, *Discourse*, p.47: 'C'est une vaste galerie, qui s'étend d'Orient en Occident, du Midi au Septentrion. Tantôt notre esprit se plait à suivre

Erudition thus founds autonomous thinking if one cultivates a critical attitude towards past doctrines. The philosophical approach advocated here by Merian demands a free, informed and well-argued choice, consequently endorsing a central stance of eclectic philosophy. It should also be observed that he introduces here the criterion of utility, albeit without explaining it further: chosen truths should be useful for reasoning and for everyday life, as they help fight against prejudice.

These methodological requirements can be met, Merian believes, and a novel approach, in line with the strategies enumerated above, should be able to ensure the progress of this science, and help it attain the eminent position it is entitled to. It becomes clear that in his view the 'briars and thorns' covering metaphysics[28] are not inherent to this discipline but ascribable to the way it has been practised.

'True' philosophers and learned societies

On Merian's account, a novel, irenic attitude towards the other sciences – literature, natural sciences and mathematics – must accompany any philosophical enterprise. The prerequisite of a restored relationship must be the admission of the primacy and specificity of metaphysics, but this should happen within a 'republic of the wise',[29] where scholars – philosophers and scientists – work together in a harmonious and fruitful manner. Philosophy, he argues, should be practised within learned societies or academies (*sociétés savantes*). This conviction, already advocated by Thomasius, is shared by other members of the Academy, like Maupertuis and Formey, but not acknowledged in contemporary accounts of eclectic philosophy. It becomes a central element of Merian's eclectic approach.

Merian's conviction that 'a new day' is rising for metaphysics also relies on several favourable circumstances: an increased autonomy with regard to theology and a supportive attitude from political rulers who protect metaphysics and guarantee its freedom. He is of

---

l'ordonnance de ses tableaux, tantôt il aime à s'égarer dans leur variété. Et que pourroit-il y avoir pour lui de plus intéressant? C'est sa propre histoire; c'est le fidelle miroir où il peut contempler sa petitesse & sa grandeur, son élévation & ses chutes.'

28. Merian, *Discourse*, p.34. Formey uses the same words to describe 'the metaphysics of the scholastics'; see Jean H. S. Formey, 'Préface', in *Histoire de l'Académie royale des sciences et des belles lettres de Berlin* (1745; Berlin, Haude, 1746), sig.)( )(2.

29. Merian, *Discourse*, p.40.

course referring here to Frederick the Great, the 'philosopher king' who had endowed the Berlin Academy with a class of speculative philosophy in the 1740s, the first in Europe, and had managed to attract many important philosophers of the time, as both full and corresponding members of the Academy. Frederick indeed envisaged this institution as defined by the collaboration between the experimental sciences and speculative philosophy, as well as by the collaboration of all its classes which met and presented their work in common and general sessions. Even the president of the Academy had been chosen by the king according to this desideratum: in the 1740s Maupertuis was both a world-renowned scientist and a philosopher. The *Discourse* ends with the injunction to the members of the Academy to take care and advance metaphysics as a joint work: its destiny depends on learned societies and not on isolated thinkers, Merian claims. Therefore, trust and optimism are justified, and progress may be envisaged. When properly conducted, this science has already proven its richness and fecundity: 'metaphysics has more than once seen its deserts transformed into charming lands when cultivated by skilful hands.'[30]

Nevertheless, the status of the 'perfect metaphysician' as imagined by systematic philosophers may well not be attainable. The abovementioned attitude towards truth endorsed by Merian denies the possibility of knowing 'in detail the vast plan' of God or believing that one day there will be no mystery, no veil, no miracle for us in the universe, as this type of thinker would like to hope.[31] Such immoderate ambition is anyway incompatible with the methodological precondition of acknowledging the limits of our knowledge even with regard to metaphysical questions, as is the wish to attain and possess the whole truth by oneself, as an isolated thinker, separated from one's peers. Merian is undoubtedly ironic when he portrays the 'perfect metaphysician',[32] and therefore one might wonder if the 'praise' he announced is not in fact a 'eulogy' for metaphysics. The French word *éloge* fits equally well with both these possibilities. It might be the eulogy of the metaphysics supported by isolated sects, by system

30. Merian, *Discourse*, p.47.
31. This critical attitude might echo Thomasius' hostility to scholastic metaphysics. The same tone is to be found in Formey, 'Préface', sig.)()(2. On the Thomasian 'attack' on metaphysical scholasticism, see Ian Hunter, *Rival Enlightenments* (Cambridge, 2001), p.209-23.
32. Merian, *Discourse*, p.14.

builders and their epigones, but not of metaphysics as it should be done. 'True' philosophy (*la vraie philosophie*) and 'true' philosophers are undoubtedly possible for Merian – one can achieve this stature by acquiring the philosophical spirit already mentioned:

> But the true philosopher avoids these pitfalls. He calmly navigates on this stormy sea. With an indifferent eye he witnesses the birth and death of opinions. He takes the true and the good wherever he finds them; the important names, the powerful parties never impress him enough to make him assume the false and the chimeric. Always free, always master of himself, humane, equitable, friend of peace and of virtue; transcendental sciences embellish his character instead of corrupting it. In this perfect equanimity which represents the reward and the triumph of the wise, he successively enjoys the pleasures of a social life and the delights of the contemplative life.[33]

Merian does not use here the term 'eclectic' but obviously proposes the very definition and decisive features of the eclectic philosopher: impartial, humble and serene, one who discerns and chooses the true and the good irrespective of their origin. Erudite and virtuous, eclectic philosophers reject authority, preserve freedom and think for themselves. The last paragraphs of the *Discourse* complete this genuine panegyric to eclectic philosophy by invoking the figure of Cicero as an embodiment of these features: 'In the Senate, at the rostrum, Cicero was an impetuous torrent, who dragged everything along [...]. Sometimes walking through all the sects, perceiving their spirit, and espousing none: sometimes talking to the heart, making virtue lovable, old age young, and friendship divine.'[34]

Merian will stay true to his eclectic convictions. More than thirty years later, in 1797, in an essay entitled 'Parallèle historique de nos

33. Merian, *Discourse*, p. 43-44: 'Mais le vrai philosophe évite ces écueils. Il vogue tranquillement sur cette mer orageuse. Il voit, d'un œil indifférent, naître & mourir les opinions. Il prend le vrai & le bon partout où il le trouve; les grands noms, les partis puissans ne lui imposent jamais au point de lui faire digérer le faux & le chimérique. Toujours libre, toujours maître de lui-même, humain, équitable, ami de la paix & de la vertu, les Sciences transcendantes embelissent son caractere, au-lieu de le corrompre. Dans cette parfaite égalité d'ame qui fait la récompense & le triomphe du Sage, il goûte, tour à tour, les douceurs de la vie sociale, & les agrémens de la vie contemplative.'

34. Merian, *Discourse*, p.49: 'Dans le Sénat, à la Tribune, Cicéron est un torrent impétueux, qui entraine tout dans sa course rapide [...]. Tantôt il se promene par toutes les sectes, en saisit l'esprit, & n'en épouse aucune: tantôt il parle au coeur, rend la vertu aimable, rajeunit la vieillesse, divinise l'amitié.'

deux philosophies nationales'[35] dedicated to Wolff and Kant, he addresses once more the question of academic philosophy. He looks back at the activity of the Berlin Academy and explicitly integrates it within the German philosophy of the eighteenth century. Unlike in 1765, when the features of the eclectic art of philosophising were both too vague for philosophers and too obvious for readers, and were not labelled 'eclectic', the term is now assumed and explained. Merian ascribes the 'true philosophical and academic spirit' to eclectic philosophy and mentions as its central characteristic the freedom to think and to express one's thought. Opposing it to the 'tribal' and sectarian attitude of both the past Wolffian school and the successful Kantianism of his time, he considers eclecticism the only 'sect or [rather] non-sect' appropriate for an academy:

> I would dare to affirm that this same eclecticism which has filled, in a certain manner, the interval between Wolff and Kant, has flowed in great part from us, or at least has been greatly encouraged by our philosophers. It reigned in their class and it is the only sect or [rather] non-sect, that ought to breathe in an Academy.[36]

On his account, it is the main merit of the Berlin Academy and especially of its speculative philosophy class to have encouraged the desire to think for oneself and to have offered the opportunity to express one's ideas in an open-minded and enlightened environment. The academic spirit has thus amply proven, he claims, its capacity to counterbalance the propensity to conflict displayed by past and present system-building metaphysicians. Merian notes that harmony, respect and peaceful collaboration have reigned within the Academy despite the different sensibilities of the members, and mentions the eclectic methodological principles that have guided its philosophical endeavour: the rejection of authority, the condemnation of systems, autonomous and free thinking, critical choice, the privilege to philosophise within a learned society. The freedom of thought and of choice, essential to eclecticism, is here associated with the possibility

35. Merian, 'Parallèle historique de nos deux philosophies nationales', in *Mémoires de l'Académie royale des sciences et belles lettres* (1797; Berlin, Decker, 1800), p.53-96.
36. Merian, 'Parallèle', p.95: 'J'oserois encore affirmer que ce même Eclecticisme qui a rempli, en quelque façon, l'intervalle entre Wolf & Kant, a coulé en grande partie de chez nous, ou du moins a été fortement encouragé par nos philosophes: il régnoit dans leur classe; & c'est la seule secte ou non-secte qui doit respirer dans une Académie.'

of submitting to the facts and to adapting and revising one's ideas. On his account, eclectic philosophers alone are always 'masters' of their philosophical standpoint and are guided by what is true and good: '[T]he philosopher who chooses, because this is what the word "eclectic" means, always remains the master of his choice, he can adjust and readjust according to the circumstances; on the contrary, systematic concatenation rules out this flexibility; its stiffness resists it; all or nothing, either to endure or to perish, this is its slogan.'[37]

He clearly states here the provisional character of the eclectic approach, which may evolve according to legitimate objections and new findings. It is not a matter of opportunism, because the eclectic philosopher ignores or is supposed to ignore authority and fame, but a case of falsification.[38] Our limited intellect is likely to err, it may have to recognise its errors and adjust its ideas, and must accept it will never attain absolute wisdom nor cement knowledge into a system, but at best will progress step by step. This is why, according to Merian, the true art of philosophising is and always will be eclectic, notwithstanding those who denigrate it.

In his remarkable work *Die wahre Aufklärung*, Werner Schneiders briefly mentions a 'new critical eclecticism' (*kritische Eklektik*) that rises in the aftermath of Wolff's death.[39] Even if this term is somewhat pleonastic if put in a Thomasian perspective where 'eclecticism' is by definition critical, it could be applied to Merian's enterprise, and has the merit of differentiating eclectic philosophy from syncretism. Schneiders discerns the influence of English thought in its elaboration and situates its main characteristic in a new moral and practical orientation in philosophy. I believe that, although not *praxisbezogen*, Merian's *Discourse on metaphysics* and its philosophical programme may be considered as delineating a theoretical form of critical

37. Merian, 'Parallèle, p.91: '[L]e philosophe qui choisit, car c'est ce que signifie le mot d'Eclectique, demeure toujours le maître de son choix, il peut se plier & se replier selon les occurrences; au lieu que la concaténation systématique exclut cette flexibilité; sa roideur y résiste; tout ou rien, durer ou rompre, voilà sa devise.' He justifies the 'title of Eclectics' by describing how these thinkers acted: 'they took everywhere, and indiscriminately accepted what seemed true and well found, without deferring to names and authorities, without becoming the clients or satellites of anyone'; Merian, 'Parallèle', p.91.
38. On the provisional status of Thomasius' philosophical position, see Holzhey, 'Philosophie als Eklektik', p.28.
39. Werner Schneiders, *Die wahre Aufklärung: zum Selbstverständnis der deutschen Aufklärung* (Freiburg and Munich, 1974), p.15.

eclecticism. Indeed, other significant members of the Berlin Academy endorsed several of its core features as well.

Academic philosophy in Formey and Maupertuis

During the same period, a similar perspective and related methodological concerns may be found in the writings of J. H. S. Formey, the permanent secretary of the Berlin Academy, and particularly in his two-part essay presented at the Academy in 1767 and 1768, 'Reflections on what could nowadays be regarded as the main goal of academies, and as their greatest benefit' ('Considérations sur ce qu'on peut regarder aujourd'hui comme le but principal des académies, et comme leur effet le plus avantageux').[40] It is noteworthy that Formey, who alongside J. G. Sulzer was for a long time the main philosopher to endorse an explicit Wolffian standpoint and provide a counterpoint to Maupertuis's views, defends in the mid-1760s a conception similar to that of Merian. This fact further supports our claim that by then the Berlin Academy had achieved a rather homogeneous philosophical position and its members agreed, if not on doctrinal content, at least on a particular set of methodological principles. They shared the conviction that a specific philosophical attitude or art of philosophising should be cultivated and promoted by this institution, which was endowed, in their view, with a specific philosophical task. Formey employs here the same key concept as Merian, namely, the 'philosophical spirit' (*esprit philosophique*), in order to designate the (only) 'true' philosophical stance.

At the very beginning of the first tract, Formey quotes Thomasius and supports the rehabilitation of Descartes. He argues that the latter had initiated a first revolution in philosophy by denouncing error and proposing novel maxims and ways to reason. Alongside Newton and Leibniz, Descartes shattered 'false-knowledge' (*le faux savoir*) and freed philosophy.[41] On Formey's account, a second revolution,

40. Jean H. S. Formey, 'Considérations sur ce qu'on peut regarder aujourd'hui comme le but principal des académies, et comme leur effet le plus avantageux', in *Histoire de l'Académie royale des sciences et belles lettres* (1767; Berlin, Haude and Spener, 1769), p.367-81 (hereafter 'Considérations' I); and 'Considérations sur ce qu'on peut regarder aujourd'hui comme le but principal des académies, et comme leur effet le plus avantageux: second discours', in *Histoire de l'Académie royale des sciences et belles lettres* (1768; Berlin, Haude and Spener, 1770), p.357-66 (hereafter 'Considérations' II).

41. Formey, 'Considérations' II, p.361.

more important than the first, was brought about by Descartes's own mistakes and blunders. His disciples, who had become servile devotees despite Descartes's clear warnings against this attitude,[42] eventually understood that even their master was wrong in some respects and that no philosopher should be followed blindly. In the second essay, Formey seems to identify these first two revolutions with a two-step attack against 'false knowledge' that makes possible a genuine philosophical approach, even if its proper articulation still lies a long way ahead. In a sombre tone, Formey condemns the eighteenth century as removed not only from true knowledge but also from an authentic philosophical attitude, despite the consensus of calling it the 'century of philosophy'.[43] On Formey's account, the liberation of the human spirit from the power of sects is not yet complete. Using a typical Thomasian trope for eclectic philosophy, he deplores the fact that the number of those able to 'see with their own eyes'[44] is very limited. Nevertheless, significant progress has been made in this respect, and sects no longer have a monopoly on philosophical knowledge.[45] No despotic privilege may be invoked any longer because nobody 'feels like' looking through somebody else's eyes or surrendering to authority alone. Thinkers desire to think for themselves, to judge and reason autonomously, and the servile attitude is being abandoned.

Nevertheless, there remains a further revolution to be carried out, according to Formey, and it consists in the banishment of what he calls 'half-knowledge' (*le demi-savoir*).[46] He considers this third revolution to be primarily the task of the academies. Formey defines half-knowledge not as a partial or incomplete knowledge, as might be expected, but as the pretence of knowing everything when in fact

42. Formey here uses concepts employed in Merian's *Discourse* and mentions the 'servile character' of the Cartesians; 'Considérations' I, p.373. Similar claims are made at that time, for instance, by Jean D'Alembert in the *Preliminary discourse to the Encyclopedia of Diderot*, translated by Richard N. Schwab (Chicago, IL, 1995), p.78-80, and in Emilie Du Châtelet's *Institutions de physique* (Paris, Prault fils, 1740), p.5-6.
43. Formey, 'Considérations' II, p.361. On the eighteenth century as the 'century of philosophy', see also Jean D'Alembert, *Essai sur les éléments de philosophie* (Hildesheim, 2003), p.9.
44. On the reception of this Thomasian image in Brucker, see Ulrich J. Schneider, 'The problem of eclecticism in the history of philosophy', *Intellectual history review* 26:1 (2016), p.117-29 (122).
45. Formey, 'Considérations' I, p.374.
46. Formey, 'Considérations' II, p.359.

one knows only a part, or, at best, the presumption of knowing more than one actually does. The illusion of universal knowledge and the impression of possessing vaster knowledge than is the case define, in his view, the most prejudicial and harmful posture for any thinker. Not only is absolute knowledge humanly unachievable, but, as Merian had also warned, this deceitful claim has devastating consequences: scholars despise each other, take their own field to be universal, try to subordinate other sciences and initiate conflicts.

Formey's definition of the 'true philosopher' as opposed to the 'half-philosopher' comprises similar elements to Merian's viewpoint: 'no bright mind, no true philosopher bears the name of any master or of any sect; but, after having thoroughly examined and carefully compared all doctrines, he adopts one because he considers it true, or he elaborates one by putting together the consistent [elements] he has found throughout his studies, and by means of his own investigations.'[47]

We consequently find in Formey comparable methodological principles: the denunciation of authority; the rejection of sects; the need to be widely read in philosophy, to thoroughly examine and compare already existing doctrines; the imperative to choose or form one's doctrine according to one's own critical judgement; the necessity to think for oneself; the attachment to truth and the rejection of prejudice; the necessity to overcome moral flaws like pride and self-sufficiency. What might be highlighted here, however, is the possibility of building or choosing one philosophical doctrine as such and adhering to it. Formey, a long-term admirer of Wolff's works, envisages this element as belonging to the philosophical spirit, whereas Merian inclines towards the development of one's own philosophical perspective by endorsing ideas taken from other doctrines without espousing an already existing systematic conception as such. Accordingly, the explicit critique of philosophical systems in general, advocated by Merian, is absent from Formey's texts.

The *esprit philosophique* alone, Formey argues, makes one a true scientist or a philosopher in a broad sense, and this title does not depend on any particular philosophical conception or doctrinal content. Nevertheless, as in Merian's case, an underlying metaphysical

47. Formey, 'Considérations' I, p.373: 'tout bon esprit, tout vrai Philosophe, ne porte plus le nom d'aucun Maître, d'aucune Secte; mais, après avoir suffisamment examiné, mûrement comparé, toutes les doctrines, en adopte une, parce qu'il la trouve vraye, ou s'en forme une en réunissant ce qu'il a trouvé de solide dans le cours de toutes ses études, & par la voye de ses propres recherches.'

stance grounds and unifies the above-mentioned set of methodological principles: namely, the thesis that ontology or general metaphysics is the 'first science' (*science première*) and should be regarded as foundational and playing a guiding role; its 'rights', he claims, must be unanimously acknowledged.[48] It is noteworthy that Formey defends a hierarchical structure comparable to Merian's: the primacy of metaphysics, which provides the principles of all the sciences, is the underlying theoretical conviction upon which the philosophical spirit can be developed. As his text has different goals, his analysis is rather succinct with respect to the definition of metaphysics.

Sharing Merian's standpoint, Formey argues that the new revolution directed against 'half-knowledge' cannot be completed by isolated thinkers, nor can the philosophical spirit be achieved on one's own. Rather pessimistic about his century, he believes that solitary true philosophers or 'apostles of reason' would not be accepted or followed by their contemporaries: they would become the victim of many evils ranging from superstition and excessive adulation to frivolity, incomprehension and hatred. This is why, on his account, the responsibility for denouncing 'half-knowledge' and cultivating the philosophical spirit falls upon learned societies, specifically upon academies alone:

> If this is feasible, it is reserved to societies, to companies alone. [...] It befalls academies to ensure the reign of a purified [*épuré*], solid knowledge, rich in precious fruits, that, so to speak, chases away half-knowledge, as it has previously chased away false knowledge. [...] Academies do not have a more essential duty nor a more glorious task to accomplish.[49]

He thus radicalises Merian's thesis and assigns this mission to academies only, which ideally should be composed of enlightened scholars of good will who spurn conflict, jealousy, partiality, ambition and personal interests. Even if this point is not addressed as such, the symphony of sciences and the harmonious collaboration of scientists

48. Formey, 'Considérations' II, p.361.
49. Formey, 'Considérations' II, p.364: 'Si la chose est faisable, ce n'est qu'à des Corps, à des Compagnies, qu'elle est réservée. [...] c'est aux Académies à faire régner un savoir épuré, solide, fécond en fruits précieux, qui donne, pour ainsi dire, la chasse au demi-savoir, comme on l'a donné précédemment au faux savoir. [...] Les Académies n'ont point de devoir plus essentiel à remplir, de tâche plus glorieuse à exécuter.'

are in fact presupposed in the model he articulates. Fifty years of such a wise academic 'dictatorship', Formey believes, would prove of great benefit to humankind in general.[50]

Both Merian and Formey endorse and develop in their essays not only Frederick's ambition to create an academy where all sciences and especially the classes of speculative and experimental philosophy collaborate, and where English and French enlightened ideas infuse the German metaphysical approach so as to forge a novel philosophical synthesis, but above all the explicit programme envisaged by Maupertuis for the Academy over which he was presiding. Indeed, in 1750, in a famous address entitled 'On the duties of the academician' ('Des devoirs de l'académicien'),[51] he defines the 'academic spirit' and invites the members of the different classes to work together. In his view, no scientist can be self-sufficient, and mutual help brings along numerous advantages like innovative ideas, quick answers, efficiency and productivity, to name but a few. Moreover, the Berlin Academy has the unique privilege of being endowed with a speculative philosophy class, Maupertuis argues, which opens the largest scientific field to its members.

The highest gain to be obtained within an academy, and only within it, is according to Maupertuis the 'academic spirit', 'the sentiment of truth' or the 'philosophical spirit'.[52] We may note that both Merian and Formey would later use the same concept. At first, Maupertuis defines it rather briefly as a sentiment that makes one recognise the truth when one sees it, and prevents one from looking for it where there is no chance of finding it. He later gives further details and describes the philosophical spirit as conferring the ability to distinguish between various degrees of assent, to differentiate evidence from probability and doubt, and to discern which status to confer to one's own ideas.[53]

The best example of the necessary collaboration between the sciences is the cooperation between experimental philosophy and speculative philosophy. The first is founded on experience, Maupertuis claims, but, without the help of speculation, the experimental

50. Formey, 'Considérations' II, p.366.
51. Maupertuis, 'Des devoirs de l'académicien', in *Œuvres de Maupertuis*, 4 vols (Lyon, Bruyset, 1768), vol.3, p.283-302 (hereafter 'Duties'). See the appendix to this volume for the translation.
52. Maupertuis, 'Duties', p.285 and 295.
53. Maupertuis, 'Duties', p.295.

philosopher wastes much time and energy, and risks getting lost in the multitude of observations and perceptions that should be ordered, understood and interpreted with the help of reasoning (*raisonnement*). By contrast, speculation requires confirmation from experience or else it might build fantasies disconnected from reality.[54] Without endorsing the primacy of metaphysics,[55] as Merian and Formey do, he nevertheless makes the case for a specific philosophical method, which neither imitates the geometrical method nor relies on the natural sciences, but takes into account the specificity of philosophy understood precisely as Merian does in the *Discourse*.

We may thus observe that the project outlined by Maupertuis in this discourse delineates the main directions which will later appear in Merian's and Formey's accounts of the philosophical spirit and proper academic philosophy. Maupertuis's programme already argues for an eclectic approach to philosophy and to truth in general, for a synthesis between the experimental and the speculative sciences, for the specificity of philosophy with respect to mathematics and the natural sciences, for an indispensable cooperation among scientists, as well as for the decisive role to be played by the Academy. It is from the outset understood and endorsed as the Academy's particular approach to philosophy, even if it would take about fifteen years for this programme to mature and to be clearly articulated and applied as a precise art of philosophising in Merian's *Discourse*.

The Thomasian legacy

It is one of the main claims of this chapter that academic eclectic philosophy, as defined by Merian and promoted by the Berlin Academy 'between Wolff and Kant' (that is, as Merian claims in his 'Parallèle historique', from the late 1740s up to the late 1780s), is consistent with the 'eclectic philosophy' (*Eclectische Philosophie*) advanced by Christian Thomasius. The source of one of the most significant ideas defining the German Enlightenment, Thomasius was the thinker who analysed and supported eclectic philosophy as the main, if not the only, approach able to ensure autonomous thinking. Whether it was called *Eklektik*, *Selbstdenken* or *Mündigkeit*, as Norbert Hinske has

54. Maupertuis, 'Duties', p.289.
55. For a recent analysis of Maupertuis's take on metaphysics, see Christian Leduc, 'La métaphysique de la nature à l'Académie de Berlin', *Philosophiques* 42:1 (2015), p.11-30.

convincingly argued, its content remained homogeneous throughout the century and preserved the main features theorised by Thomasius.[56] This programmatic idea was to traverse the eighteenth century and be upheld until Kant. In view of the writings of the members of the Berlin Academy, we may talk about a revival or a new flourishing of this idea after the mid-century, as we have seen in Merian and Formey. Werner Schneiders is among the few scholars to mention such a renewed interest in Thomasian philosophy around the mid-eighteenth century, but he attributes it to the emphasis on moral aspects and to the articulation of a 'philosophy of the world' (*Philosophie der Welt*) or of a 'philosophy for the world' (*Philosophie für die Welt*)[57] as opposed to the scholarly, abstract and demonstrative philosophy endorsed by the Wolffians (*Schulphilosophie*). Schneiders' standpoint is undoubtedly accurate with regard to the goals of the *Popularphilosophie*. However, I would like to highlight another, less studied, element of the Thomasian legacy and to argue here for the rehabilitation of the latter's theses on eclectic philosophy, and, moreover, for a novel and strong interest in this aspect of his thought owing to the Berlin Academy.

In his *Introductio ad philosophiam aulicam* of 1688, translated into German in 1710 as *Einleitung in die Hof-Philosophie*, Thomasius affirms that the 'most excellent philosophers have always been eclectics (*Eclectici*)'.[58] He sets them in contrast with the members of 'sects', narrowly understood, who blindly follow and defend the doctrine of a unique teacher. He also distinguishes eclectic thinkers from 'syncretists' (*Syncretisten*), who through imprudent comparisons render certain ideas uncertain, as well as from 'Pyrronists' (*Pyrrhonii*), who do not believe in anything.[59] Thomasius seems to make the case for an original point of view (*Ich nenne aber*) and to support an approach that was subject to debate at the time:

> However, I call eclectic philosophy one that requires one not to depend on the sayings of one single philosopher, nor to completely

56. Norbert Hinske, 'Die tragenden Grundideen der deutschen Aufklärung', in *Die Philosophie der deutschen Aufklärung*, ed. Raffaele Ciafardone (Stuttgart, 1990), p.407-58 (417-24).
57. Werner Schneiders, 'Zwischen Welt und Weisheit: zur Verweltlichung der Philosophie in der frühen Moderne', in *Philosophie der Aufklärung – Aufklärung der Philosophie: Gesammelte Schriften* (Berlin, 2005), p.343-64 (355).
58. Christian Thomasius, *Einleitung zur Hof-Philosophie* (Frankfurt and Leipzig, Rüdger and Bauern, 1710), I, §93, p.52.
59. Thomasius, *Hof-Philosophie*, I, §89, p.50.

> commit oneself to a unique teacher, but to collect from the sayings and writings of various teachers each and every true and good thing in the treasury of one's own understanding. [And it also requires one] not to reflect so much on the authority of the teacher, but to examine by oneself whether a certain element of doctrine is well founded, and to add something from oneself to it, and thus to look rather through one's own eyes than through someone else's.[60]

His definition of eclectic philosophy thus puts forward the requirement to think for oneself, to freely read and judge various doctrines and to choose whatever appears to be 'true and good'. Moreover, even if the term *Selbstdenken* is not mentioned here, one should add his own ideas to this selection of theses and thus not only think autonomously but also be creative. He therefore goes beyond the understanding of eclectic philosophy as a mere combination of already formulated doctrines. Rejection of authority, careful examination of already elaborated sets of principles, independence of thought and the discovery of new truths are his central imperatives. He further stresses the necessity to act impartially and fairly, and to show an 'equal love' (*gleiche Liebe*) and 'equal freedom' (*gleiche Freyheit*) towards every conception.[61] 'Very necessary' and 'very useful' in order to examine the truth, Thomasius believes, this way of philosophising neither despises past doctrines nor neglects new ones, but approaches them all with respect, interest and a critical spirit. The sectarian attitude is its opposite in every way: it is immodest, exclusivist, biased and bellicose, and should therefore be denounced.

Not only is eclectic philosophy an appropriate way to search for truth, on Thomasius' account, but it is also able to avoid conflicts and disputes. It allows all sects to express their points of view and, far from any partial and prejudiced attitude, it neither vehemently supports nor combats specific schools. He seems to particularly value this consequence. In order to define the exemplary attitude

60. Thomasius, *Hof-Philosophie*, I, §90, p.50: 'Ich nenne aber eine Eclectische Philosophie eine solche, welche da erfordert, daß man von dem Munde eines einzigen Philosophi allein nicht dependiren, oder denen Worten eines einzigen Lehr-Meisters sich mit einem Ende verpflichten soll, sondern aus dem Munde und Schrifft allerley Lehrer, alles und jedes was wahr und gut ist, in die Schatz-Kammer seines Verstandes sammlen müsse, und nicht so wohl auf die Autorität des Lehrers Reflexion mache, sondern ob dieser und jener Lehr-Punct wohl gegründet sey, selbst untersuche, auch von dem Seinigen etwas hinzu thue, und also vielmehr mit seinen eigenen Augen als mit anderen sehe.'
61. Thomasius, *Hof-Philosophie*, I, §92, p.51.

of an eclectic philosopher, Thomasius employs here the concept of *Billigkeit*,[62] translated into French as *sentiment du juste* and later used by Maupertuis and Merian in their descriptions of the 'academic' or 'philosophical spirit'. Fair-minded, irenic yet erudite and promoting critical and independent thinking as well as creativity, eclectic philosophy, also named *Wahl-Philosophie* (philosophy of choice),[63] shows the proper way to philosophise.

One decisive aspect to be highlighted here is Thomasius' emphasis on the importance of eclectic philosophy for the learned societies of his time. He considers most of them to openly embrace this art of philosophising in various European countries. He cites Germany, Italy, France, England and Denmark and specifically mentions the Royal Society and the French Academy of Sciences. He thus sees this approach at work within this *gelehrte Gesellschaften*, and articulates a particular relationship between the eclectic attitude, on the one hand, and the way of functioning and the contribution of these institutions, on the other. Their members acknowledge themselves to be followers of the eclectic way to philosophise, according to Thomasius.[64] While universities still endorse Aristotelian philosophy – the main target of Thomasius' fight against authority – and therefore remain attached to 'sects', by contrast, courts, academies and scholars are apt to, and therefore should, embrace eclecticism. The Berlin Academy, as we have seen, definitely takes up this suggestion and regards the eclectic philosophy as the only 'sect or [rather] non-sect' to be promoted by its members, as the only attitude compatible with and worthy of an academy, as Merian puts it in 1797.[65] This particular feature – the eclectic philosophy as the main task to be taken up by philosophical societies – clearly expresses a Thomasian legacy assumed by academic philosophy, and yet surprisingly is absent from contemporary descriptions of eclectic philosophy.

Responding to several objections, Thomasius recommends the practice of eclectic philosophy to mature intellectuals, but not to inexperienced young students, as it implies choice and discernment. To the question of whether a 'whole and single' philosophy should

62. Thomasius, *Hof-Philosophie*, I, §91-92, p.51.
63. On the relationship between philosophical choice and Cartesian doubt in Thomasius, see Werner Schneiders, 'Vernünftiger Zweifel und wahre Eklektik', *Studia Leibnitiana* 17:2 (1985), p.143-61 (147-48).
64. Thomasius, *Hof-Philosophie*, I, §93, p.54.
65. Merian, 'Parallèle', p.95.

be preferred to a doctrine composed of various, wavering and unstable parts, he responds that such a definitive and unchanging conception, worthy of the (pompous) name of 'true philosophy', is not attainable, owing to the difficulty of the issues involved and to the weakness of our understanding.[66] This is why we should be 'satisfied' with another philosophy, namely the eclectic one, which he compares to a serviceable ship that has been repaired and adjusted here and there. It is better to navigate on such a ship, Thomasius claims, than to be on a ship full of holes and cracks that has remained unchanged. He also employs here what would become a privileged metaphor of eclectic philosophy in the eighteenth century: eclecticism is compared to a palace built by several artists as opposed to a bad cabin erected by a single artisan. What is noteworthy here is that Thomasius excludes the possibility of assembling or articulating proper doctrinal content and adhering to it – eclectic philosophy is and must remain a work in progress, always adjustable and needing the contribution of various thinkers. It is first and foremost a methodological approach, not a fixed set of theses to be put together and adopted at some point. Thus, eclectic philosophers do not encounter the risk of eventually becoming founders or members of a specific sect themselves.

In his *Hof-Philosophie*, Thomasius criticises scholastic metaphysics, accusing it of mixing revealed and natural knowledge and subordinating general metaphysics to theology.[67] He considers rationalist metaphysics the sectarian philosophy[68] *par excellence* and elaborates his early eclectic methodology as an alternative approach to truth. Contrary to Thomasius' rejection of metaphysics, Merian and Formey, probably under the influence of Wolff's philosophical project, would apply this eclectic methodology to metaphysics as such, thus articulating a novel treatment of this science with the help of Thomasian eclectic principles.

Thomasius' views on eclectic philosophy[69] deeply influenced the eighteenth century, despite the relative shadow cast upon his standpoint by the success of Wolff's *Schulphilosophie*. The so-called

66. Thomasius, *Hof-Philosophie*, I, §98, p.54
67. Thomasius, *Hof-Philosophie*, II, §22-41, p.67-74.
68. Thomasius, *Hof-Philosophie*, II, §63, p.81.
69. On the evolution of Thomasius' views on eclectic philosophy, see Michael Albrecht, 'Thomasius – kein Eklektiker?', in *Christian Thomasius 1655-1728: Interpretationen zu Werk und Wirkung*, ed. Werner Schneiders (Hamburg, 1989), p.73-94.

'Thomasian-Pietist' tradition[70] embodied by thinkers such as J. F. Budde, J. G. Walch, Ch. A. Crusius and J. J. Brucker endorsed several central theses specific to the eclectic 'art of philosophising', and disseminated it to a larger European audience. Despite the common name given to this direction, their viewpoints are not always homogeneous, and some significant specificities are later reflected in the philosophy of the Academy.

Budde's treatment of eclecticism in his *Elementa philosophiae instrumentalis seu Institutiones philosophiae eclecticae* (1703), undoubtedly influenced by Thomasius' views, had a huge impact on the first decades of the eighteenth century. Budde endorses the 'eclectic way to philosophise' in the preface,[71] and stresses the importance of autonomously elaborated arguments based on the consideration of the nature of things. Knowledge should be based not on others' sayings, but on the things themselves.[72] While the knowledge of 'ancient and new' ideas seems to remain important,[73] it is thinking for oneself, proving one's arguments and freely verifying all received views that become decisive methodological principles. Budde advocates the elaboration of a coherent and systematic conception: self-evident principles should found the rigorous deduction of connected arguments and ensure the order and systematicity of a true philosophical doctrine. He refuses to cite names or doctrines, and claims that others' views may be admitted only if they correspond to – a seemingly already discovered – truth. While he recommends a position that is elaborated according to the eclectic method (*eclectico more*), the accent now falls on systematic deduction, on free verification and particularly on free and independent thinking. Systematic and eclectic procedures are brought together, and regarded as the result of individual, exceptional minds, not subordinated to sects and authority. These theses would strongly

70. See Sonia Carboncini, 'Die thomasianisch-pietistische Tradition und ihre Fortsetzung durch Christian August Crusius', in *Christian Thomasius*, ed. W. Schneiders, p.287-303. However, the label 'Thomasian-Pietist' is too vague and attempts to unify positions that are, in my view, too heterogeneous.

71. Johann F. Buddeus, *Elementa philosophiae instrumentalis seu Institutiones philosophiae eclecticae*, 10th edn, vol.1 (Halle, Orphanotropheum, 1755), p.xxi.

72. 'Ex rebus enim, non ex aliorum verbis aut effatis sapere, hoc demum philosophie esse censeo'; Buddeus, *Elementa*, p.xxii.

73. On Budde's eclecticism before 1703, see Michael Albrecht, *Eklektik: eine Begriffsgeschichte mit Hinweisen auf die Philosophie- und Wissenschaftsgeschichte* (Stuttgart and Bad Cannstatt, 1994), p.434-40.

influence authors such as Walch, Brucker and probably Formey, and, to a lesser extent, as we have seen, Merian.

Johann G. Walch's *Philosophisches Lexikon* (1726) was widely read and reprinted several times in those decades. Taking up significant Thomasian standpoints and explicitly quoting Budde's works, Walch devotes a special entry to 'eclectic philosophy' (*Eclectische Philosophie*).[74] He defines it as a particular art of philosophising where the thinker relies on past doctrines and conclusions that he analyses and uses according to his own reason and judgement. Subordinated to the quest for truth, these doctrines must be understood in a 'lively' (*lebhaft*) manner and ensure the discovery of novel truths. He hereby implies that, even if erudition is necessary, a mere passive 'historical' knowledge does not suffice; discernment and active, justified choice are, according to Walch, specific elements of eclectic philosophy. Opposed to sectarianism, scepticism and syncretism, eclecticism supposes freedom of thought, personal convictions, creativity and analytical thinking. Not only is independence of thought necessary, it has to be accompanied by the appropriation of already existing conceptions into a coherent and original reflection which makes possible the discovery of novel insights. The third edition of the *Lexikon*, published in 1740, adds a new entry on 'Eclectici' while leaving the paragraph about eclectic philosophy unchanged. Mainly devoted to ancient thinkers and to the eclectic treatment of Christianity, this new entry has a less enthusiastic tone. Walch restricts the label of 'eclectics' to 'certain' (*gewisse*) philosophers to whom he still grants the capacity to 'intelligently choose' the best among past theories.[75] Nevertheless, he is not convinced that this method has always had the good effect it intended, at least not in the past, and seems to be a little disoriented by the numerous contemporary thinkers ('an almost innumerable amount') presenting themselves as eclectics.[76] This may be one of the reasons why he prefers to focus on Potemo of Alexandria and his disciples, and refers to the works of Sturm, Thomasius and Budde for, perhaps, an updated view on this approach. However, the multitude of philosophers claiming to be eclectics (or having been attributed this label) is here clearly an issue for the pertinence of the concept in the 1740s.

This latter aspect will become particularly obvious in Brucker's treatment of eclecticism. Brucker devotes the second part of the fourth

74. Johann G. Walch, *Philosophisches Lexikon* (Leipzig, Gleditsch, 1726), p.835.
75. Walch, *Lexikon*, p.835.
76. Walch, *Lexikon*, p.834-35.

volume of his famous *Historia critica philosophiae* (1744) to 'eclectic philosophy', which he sees as beginning in the fifteenth and sixteenth centuries and as following the 'syncretistic' period of the Renaissance. 'Victorious' over sectarian philosophy as well as over syncretism, this philosophy has, according to him, rejected the prejudice of authority and veneration of the ancients, and has succeeded in building new doctrines from diverse elements chosen from past schools of thought. Owing to 'great and sharp' spirits, who have checked, judged and accepted only verified, demonstrated truths, novel philosophical systems have been built upon true and universal principles (*principia vera, universalia*).[77] Brucker insists here on the necessary recourse to one's own (innate) reason. He names almost all the important philosophers since Bruno and Bacon up to Wolff as 'restorers' of eclectic philosophy,[78] thus undoubtedly contributing to the dilution of the concept: not only is it difficult to grasp how these thinkers revived something that did not exist prior to them – Brucker restricts eclectic philosophy to the modern age and denies its presence in Antiquity and the Renaissance – but it also becomes problematic to conceive of a proper unity among such diverse and numerous conceptions. Nevertheless, we recover here several of the methodological elements already mentioned by Thomasius and Budde as essential for eclectic philosophy: the independence of thought, the rejection of authority and of the sectarian attitude, the justified choice of doctrinal views and the elaboration of novel conceptions. What is, however, endorsed by neither Thomasius nor Merian is the emphasis on the activity of individual thinkers who through their own, independent reflection and efforts can erect rigorous systems of well-connected elements.[79] We have already mentioned the central role played by learned societies in attaining and attesting to the 'spirit of truth' that defines eclectic philosophy for Merian (and Thomasius), as well as Merian's explicit

77. Johann J. Brucker, *Historia critica philosophiae*, vol.4.2 (Leipzig, Breitkopf, 1744), §I, p.4.
78. The first book of the second part of the fourth volume is entitled 'De restauratoribus philosophiae eclecticae universae'.
79. Brucker, *Historia*, §II, p.4. On eclectic philosophy in Brucker, see also Leo Catana, *The Historiographical concept 'system of philosophy': its origin, nature, influence and legitimacy* (Leiden, 2008), who argues for its derivation from systematic philosophy (p.26). The features listed by Catana (see p.25-34) as defining Brucker's account are precisely those discussed by Budde and prove the latter's decisive influence on the *Historia*. It is thus clear that Budde, who is not mentioned by Catana, is the source of Brucker's views on eclectic philosophy.

rejection of philosophical systems. These two divergences alone suffice to downplay Brucker's (and Budde's) role in the elaboration of the eclectic academic philosophy, and to support a more direct and an unquestionably stronger Thomasian influence on this perspective.

Volume 5 of the *Encyclopédie* (1755) reprises certain passages from Brucker's *Historia critica* and sums up his theses in Diderot's article 'Eclectisme', thus ensuring their dissemination in France. The eclectic is defined as the philosopher who rejects prejudice, tradition, universal agreement and authority, and dares to think for themselves; as the philosopher who examines everything and counts only on their own reason and experience to form a unitary conception: 'from all the philosophies he has analysed without reverence and partisanship, he makes for himself a particular and domestic[80] one which belongs to him.'[81] Opposed to sectarian and to syncretic thinkers, the eclectic wishes 'to see with his own eyes', and therefore all good philosophers may be regarded as eclectics. There is, however, a revival of ancient eclecticism in modern times, and Diderot names Bacon as its founder. The author carefully separates eclecticism from syncretism, defining the latter as 'at most an apprenticeship' of eclecticism.[82] He distinguishes an 'experimental' eclecticism (which consists in assembling known truths and given facts by studying nature) and a 'systematic' eclecticism (which compares known truths and combines given facts in order to explain phenomena).[83] According to the eclectic method, assembled truths are put together neither in an arbitrary way nor according to an *a priori* plan, but by starting from first and verified principles. All subsequent propositions should be directly deduced from these principles, or else judgement be suspended until the intermediate steps can be proven. Diderot claims to notice a new tendency to extend the application of the eclectic method to other disciplines and, like Formey later on, he encourages it.

Despite these common theses shared with Thomasius, and notwithstanding Merian's probable acquaintance with the article 'Eclectisme' from the *Encyclopédie*, several of Diderot's (and Brucker's) ideas are neither accepted nor mentioned by Merian. The extensive

80. These are Brucker's very words in the *Historia*, §1, p.4.
81. Denis Diderot, entry 'Eclectisme', in *Encyclopédie, ou Dictionnaire raisonné des sciences, des arts et des métiers*, ed. Denis Diderot and Jean D'Alembert, vol.5 (Paris, Briasson *et al.*, 1755), p.270-93 (271).
82. Diderot, 'Eclectisme', p.271.
83. Diderot, 'Eclectisme', p.284.

section devoted to syncretism, the more subtle but still apparent importance of systematic doctrines, the long and generous list of 'eclectic' thinkers comprising almost all significant philosophers are elements not endorsed by Merian. Thus, there is no compelling reason to consider this article as the main source of the philosophical stance adopted by the Berlin Academy, although, given the textual proximity to Brucker's *Historia*, it may be regarded as enabling the dissemination of a standpoint belonging to the Thomasian tradition. Thomasius' works, widely circulated at the time and quoted by academic philosophers, as well as contemporary German accounts of eclectic philosophy should, therefore, be considered the decisive influence on the eclectic philosophy promoted by the Berlin Academy, and a key reason why academic philosophy must be inscribed in the German enlightened tradition.

Conclusion

I hope to have shown in this chapter that in the early 1760s a specific approach to metaphysics is developed within the Berlin Academy. It is exemplified in Merian's *Discourse on metaphysics* and both Merian and Formey theorise its main features. This academic philosophy takes the form of a set of eclectic methodological principles founded on two main stances, the primacy and the specificity of metaphysics. The eclectic philosophy advanced within the Berlin Academy revisits and takes up central Thomasian themes, and thus contributes to the revival of the latter's thought after the mid-eighteenth century. Unlike Thomasius, however, who is critical of metaphysics, it grants a central, foundational role to this science. Even if usually disregarded in the analyses devoted to post-Wolffian eclecticism, the Berlin academic philosophy indisputably advocates an eclectic approach. Moreover, it elaborates a specific and autonomous perspective, distinct both from the *Frühaufklärung* and the contemporary *Popularphilosophie* on account of its commitment to metaphysics, focus on theory and emphasis on the crucial philosophical role played by learned societies. Rooted in the early German enlightened tradition through the Thomasian methodological programme which it continues and promotes, the eclectic philosophy of the Berlin Academy thus undeniably belongs to the *Aufklärung*, or, in Dilthey's words: 'this Academy was and remained the instrument [*Organ*] of the German Enlightenment.'[84]

84. Dilthey, 'Friedrich der Grosse', p.142.

In search of a 'golden mean': experience and reasoning according to Louis de Beausobre

ANGELA FERRARO

Preliminary remarks

Louis de Beausobre (1730-1783), who joined the Berlin Academy in 1755, has met with scholarly interest only occasionally, and this owing to his participation in the scepticism debate[1] and, to a lesser extent, a few of his memoirs, especially on madness.[2] Even if they have the merit of having saved Beausobre from scholarly oblivion, the few existing publications dedicated to his work nevertheless do not get to the heart of his thinking. This chapter, I hope, will contribute to filling that void by focusing on a fundamental problem, the role of experience and its relationship with reasoning, which preoccupied

1. See Sébastien Charles, 'Des excès dogmatiques à la guérison sceptique: le pyrrhonisme raisonnable de Louis de Beausobre', *Libertinage et philosophie au XVIIe siècle* 12 (2010), p.205-17, and Sébastien Charles, 'Lumières allemandes et scepticisme: le cas de l'Académie de Berlin', in *Orthodoxie et hétérodoxie: libertinage et religion en Europe au temps des Lumières*, ed. Marie-Hélène Quéval (Saint-Etienne, 2010), p.139-49; John Christian Laursen, 'Tame skeptics at the Prussian Academy', *Libertinage et philosophie au XVIIe siècle* 12 (2010), p.219-28; Nicolas Correard, 'Reasonable scepticism in the French Enlightenment: some connections between Jean-Baptiste Boyer d'Argens, Louis de Beausobre, and Voltaire', in *Scepticism in the eighteenth century: Enlightenment, Lumières, Aufklärung*, ed. Sébastien Charles and Plínio J. Smith (Dordrecht, 2013), p.173-88; Anton M. Matytsin, *The Specter of skepticism in the age of Enlightenment* (Baltimore, MD, 2016).
2. Franco Crispini, 'La psicologia wolffiana in Samuel Formey e Louis de Beausobre', *Rivista di storia della filosofia* 44:3 (1989), p.477-97; Olivia Pallenberg, 'Immaginazione e follia: una discussione all'Accademia di Berlino', in *Dal Cartesianismo all'Illuminismo radicale*, ed. Carlo Borghero and Claudio Buccolini (Florence, 2010), p.143-58.

Beausobre throughout his career and runs like a guiding thread through his otherwise rather heterogeneous philosophical writings.[3]

There is, in Beausobre, a sustained interest in the model of cooperation between experience and reasoning that Christian Wolff had theorised in terms of both natural philosophy and psychology.[4] In his *Preliminary discourse on philosophy in general* (*Discursus praeliminaris de philosophia in genere*) of 1728, the German philosopher had divided natural philosophy into two parts, one experimental, the other dogmatic, and posited a parallel division in psychology, which, according to him, consists of a rational and an empirical part.[5] He had also created the conditions for setting in motion a sort of virtuous circle between the two branches of each discipline, that is, between the dogmatic and the experimental part of natural philosophy and the rational and the empirical part of psychology. As he states with admirable clarity in his 1732 *Empirical psychology* (*Psychologia empirica*): 'The psychologist imitates the astronomer who infers [*eruit*] a theory from observations and in turn proves the theory he infers by observations, and thanks to theory and observations he deduces other things that would have never crossed his mind otherwise.'[6]

It is within this framework of a Wolffian philosophical culture that Beausobre develops his thinking and forges the tools necessary to position himself in contemporary debates.[7] In the three parts of my

3. For the life and work of Beausobre, see John Christian Laursen, 'Louis Isaac de Beausobre (1730-1783)', in *The Bloomsbury dictionary of eighteenth-century German philosophers*, ed. Heiner F. Klemme and Manfred Kuehn (London, 2016), p.50-52.
4. See Jean Ecole, 'Des rapports de l'expérience et de la raison dans l'analyse de l'âme ou la *Psychologia empirica* de Christian Wolff', *Giornale di metafisica* 21:4-5 (1966), p.589-617, and Anna Maria Vittadello, 'Expérience et raison dans la psychologie de Christian Wolff', *Revue philosophique de Louvain* 71 (1973), p.488-511.
5. Christian Wolff, *Philosophia rationalis, sive Logica* (Frankfurt, Renger, 1728), §107-12, p.48-51.
6. Wolff, *Psychologia empirica* (Frankfurt, Renger, 1732), §5, p.4. All translations are my own unless otherwise stated.
7. Beausobre had attended Alexander Gottlieb Baumgarten's lessons and began reading Wolff under the aegis of Jean Henri Samuel Formey, as we learn from his correspondence with Formey himself: 'Monsieur Baumgarten nous lit le Droit de nature, dommage qu'il est trop diffus, si cela continue il nous faudra 2 ans pour achever ce cours. On ne lit ni métaphysique, ni physique ce semestre-ci, je n'ai pû me résoudre à écouter la logique de Mr Baumgarten: cependant, pour ne point perdre de vue une science sans le secours de la qu'elle on étudie en vain, j'ai pris la résolution de lire le grand ouvrage de Wolf avec l'aide du Seigneur: voudriez-vous bien, Monsieur, m'honorer de vos conseils à

chapter, I hope to shed light on Beausobre's relationship to his primary inspiration and to his contemporary interlocutors. These three parts correspond to three contexts in which Beausobre articulates his views on the experience–reasoning relationship. I will begin with his natural philosophy, dwelling on the *Philosophical dissertation on the nature of fire* (*Dissertation philosophique sur la nature du feu*) of 1753, and then move to his more general methodological considerations in the *Philosophical dissertation on the different parts of philosophy and mathematics* (*Dissertation philosophique sur les différentes parties de la philosophie, et des mathématiques*) and in *Reasonable Pyrrhonism* (*Pyrrhonisme raisonnable*), published in 1753 and 1755 respectively. In the third and final part, I will turn to the psychology Beausobre presents in his Academy contributions from the 1750s and 1760s.

The test case of natural philosophy

As we shall see, it is in this latter domain, psychology, that Beausobre applies the Wolffian paradigm in the most original manner. Chronologically, however, he first has occasion to reflect on the modalities of employing empirical and rational tools in studying natural philosophy. Thus we read in the *Philosophical dissertation on the nature of fire*:

> Let us not too lightly draw conclusions from the experiments we see: there are too many mistakes of omission and *subreption* that slip into our reasoning. We do not see everything. In philosophy, experience is not *the only light of truth* [...]. We have to reason, to observe nature, and to conduct experiments ourselves. In natural philosophy, to resign oneself to one or other of these three means is to construct systems which are either imaginary, founded only on abstract principles, or quite disjointed, inconsequential.[8]

---

cet égard et faire partir de votre cabinet un rayon de vos lumières pour éclairer les ténèbres du lieu que j'habite'; Beausobre to Formey (29 April 1749), quoted in Anne Baillot, 'Pyrrhonismus und Politik an der Akademie der Wissenschaften um 1750', in *Französisch-deutsche Kulturräume um 1800. Bildungsnetzwerke – Vermittlerpersönlichkeiten – Wissenstransfer*, ed. Anna Busch, Nana Hengelhaupt and Alix Winter (Berlin, 2012), p.25-42 (32). On Beausobre's correspondence, see Anne Baillot, 'Louis de Beausobre, entre cour et académie: la correspondance d'un intellectuel francophone en Prusse au XVIIIe siècle', in *Entrer en communication de l'âge classique aux Lumières*, ed. Pierre-Yves Beaurepaire and Héloïse Hermant (Paris, 2013), p.123-40.

8. Louis de Beausobre, *Dissertations philosophiques dont la première roule sur la nature du feu et la seconde sur les différentes parties de la philosophie, et des mathématiques*

To support the idea that a combined and correct employment of both experience and reasoning is necessary, Beausobre here makes use of two elements of Wolffian theory: the concept of a *fault of subreption* (*vitium subreptionis*) and the articulation of experience, in natural philosophy, between *observation* (*observatio*) and *experiment* (*experimentum*). Wolff had indeed warned against one of the major risks of confining oneself to experience alone: 'We call the *fault of subreption* the error committed in the act of experiencing when we seem to experience for ourselves what we experience least of all.'[9] This fault is committed, for example, by those who think they experience the physical influence of the soul on the body, the attractive force of magnets, or love and hate belonging to inanimate objects. The same fault introduced many chimeras into scholastic natural philosophy.[10] Wolff had also shown that our *a posteriori* knowledge (or through experience) derives from two sources: *observation* (*observatio*), which is the experience of what nature does without our intervention, and *experiment* (*experimentum*), which is the experience of what nature does as a result of our intervention.[11]

The *Philosophical dissertation on the nature of fire* also provided Beausobre with a first opportunity to wade into the scientific debates of his time. He openly criticised the standpoint of Emilie Du Châtelet, who had subscribed to the views of Herman Boerhaave and considered

(Paris, Durand and Pissot, 1753), p.96-98 (original emphasis): 'Gardons nous donc de tirer à la légere des conséquences des expériences que nous voyons: il y a trop de ces fautes d'omission & de *subreption* qui se glissent dans nos raisonnemens. Nous ne voyons pas tout. L'expérience n'est pas en Philosophie *le seul flambeau de la vérité* [...]. Il faut raisonner, observer la nature, & faire soi-même des expériences: se contenter de l'un ou de l'autre de ces trois moyens dans la Physique, c'est s'exposer ou à bâtir des systèmes d'imagination fondés uniquement sur des principes abstraits, ou bien en faire de peu liés, d'inconséquens.'

9. Wolff, *Logica*, §668, p.484 (original emphasis): 'Errorem in experiundo commissum, quo nobis videmur experiri, quod minime experimur, dicimus *Vitium subreptionis in experiundo*.'
10. See also Wolff, *Logica*, §668, p.484: 'Vitium hoc committunt, qui sibi experiri videntur influxum physicum animæ in corpus, vim attrattricem magnetis, amorem et odium rerum inanimatarum. Multas idem chimæras in physicam scholasticam invexit.'
11. Wolff, *Psychologia empirica*, §456, p.357: '*Observatio* est experientia, quæ versatur circa facta naturæ sine nostra opera contingentia. *Experimentum* est experientia, quæ versatur circa facta naturæ, quæ nonnisi interveniente opera nostra contingunt.'

rarefaction, rather than heat and light, as signalling the presence of fire or being its constant effect.[12] In her *Dissertation on the nature and propagation of fire* (*Dissertation sur la nature et la propagation du feu*), which was written in 1739 as a prize essay for the Paris Academy of Sciences and published immediately in the memoirs of the Academy, Du Châtelet had set out to verify whether fire is always hot and luminous. She concluded that several experiments convince us that it is not and that the most universal effect of fire, its distinctive property, is rarefaction.[13] Beausobre for his part objected to the idea that it was possible to resign oneself to empirically observing associations and dissociations of phenomena alone: by stopping there, it was impossible to be sure that no faults of subreption crept into our reasoning, as counterexamples could always be found. Du Châtelet, to whom Beausobre refers as a 'woman philosopher' (*femme philosophe*), was therefore wrong to consider experience the 'torch of truth' (*flambeau de la vérité*): experience alone is not enough, it must always be accompanied by reasoning.[14]

Beausobre certainly had in mind a passage from Du Châtelet's 1740 *Foundations of natural philosophy* (*Institutions de physique*, renamed *Institutions physiques* in the 1742 edition), which insists on the importance of experience:

> Remember, my son, in all your studies, that experiment is the staff that nature has given to us, blind as we are, to guide us in our research; with its help we do not stop going far, but we cannot avoid falling if we leave it out; experiment lets us know physical qualities, and it is up to our reason to use them and to draw new knowledge and new light from them.[15]

12. Herman Boerhaave, *Elementa chemiæ*, vol.1 (Leiden, Isaacum Severinum, 1732), 'De igne', p.126-424. On the influence of Boerhaave's views, see Hélène Metzger, 'La théorie du feu d'après Boerhaave', *Revue philosophique de la France et de l'étranger* 109 (1930), p.253-85.
13. Emilie Du Châtelet, *Dissertation sur la nature et la propagation du feu* (Paris, Prault fils, 1744), p.3-4 and 8-10.
14. Beausobre, *Dissertations philosophiques*, p.202-203.
15. Emilie Du Châtelet, *Institutions de physique* (Paris, Prault fils, 1740), p.10: 'Souvenez-vous, mon fils, dans toutes vos Etudes, que l'Expérience est le bâton que la nature a donné à nous autres aveugles, pour nous conduire dans nos recherches; nous ne laissons pas avec son secours de faire bien du chemin, mais nous ne pouvons manquer de tomber si nous cessons de nous en servir; c'est à l'Expérience à nous faire connaître les qualités Physiques, & c'est à notre raison à en faire usage & à en tirer de nouvelles connaissances & de nouvelles lumieres.'

However, the reproach that, in her scientific practice, Du Châtelet naively limited herself to experience is unfair and hardly justified. The ending of the quotation is enough to establish this. Beausobre must have known that the author of the *Foundations of natural philosophy* had not only highlighted the role of experience, but also insisted on the importance and usefulness of abstract principles, and of the Principle of sufficient reason in particular.[16] As she had noted, Gottfried Wilhelm Leibniz, in elaborating the principle and introducing it into scientific practice, had done a great service to the sciences since, 'for the most part, false reasoning has no other source than forgetting about sufficient reason.'[17] The Principle of sufficient reason acts as a bridle for the imagination and is 'the only thread which can guide us in the labyrinth of errors the human mind has built for itself'.[18] Nothing that violates it must be admitted. Those who deny this principle resemble 'the inhabitants of fantastic, non-existent dream worlds'.[19] Ultimately, then, Du Châtelet's views were not as incompatible with Beausobre's as he suggested. This is even more evident when we consider that the one abstract principle whose usefulness in natural philosophy he explicitly affirms is, precisely, the Principle of sufficient reason. If we were lacking this principle, he writes in the *Dissertation*, 'it would be impossible for us to demonstrate quite a few truths.'[20] He also provides examples to show that the Principle of sufficient reason provides the foundation for most demonstrations; thanks to it, Leibniz had been able to show the impossibility of the widely accepted hypothesis of absolute space and to demonstrate that space is but a relation established by discrete objects that exist simultaneously. It had been equally useful in demonstrating the impossibility of a vacuum in the universe.[21]

16. On Du Châtelet's method, see *Emilie Du Châtelet between Leibniz and Newton*, ed. Ruth E. Hagengruber (Dordrecht, 2012), and Anne-Lise Rey, 'Le leibnizo-newtonianisme: la construction d'une philosophie naturelle complexe dans la première moitié du 18e siècle: la méthode d'Emilie Du Châtelet entre hypothèses et expériences', *Dix-huitième siècle* 45:1 (2013), p.115-29.
17. Du Châtelet, *Institutions de physique*, p.27.
18. Du Châtelet, *Institutions de physique*, p.27.
19. Du Châtelet, *Institutions de physique*, p.26.
20. Beausobre, *Dissertations philosophiques*, p.188.
21. Beausobre, *Dissertations philosophiques*, p.188-91.

Towards a wise methodology

Given this basic agreement between the two authors, what other reasons are there for the reproach Beausobre addressed to Du Châtelet and her conception of experience? The fact that Julien Offray de La Mettrie cited and praised the passage from the *Foundations of natural philosophy* cited earlier may have played a role in this story. In the 1751 *Abstract of philosophical systems* (*Abrégé des systêmes*), La Mettrie's criticism of Wolffian thought concludes by mentioning Du Châtelet's appraisal of experience.[22] We may thus suppose that Beausobre, a staunch Wolffian and opponent of the materialists, sought to react less to Du Châtelet and her use of experience than to La Mettrie and his positions.[23] We may also suppose that the critique of the philosophical systems put forward by La Mettrie contributed to Beausobre's resolve, in *Philosophical dissertation on the different parts of philosophy*, to speak out against criticisms of principles and abstract systems. Yet this time, too, the official target of his criticism is not La Mettrie but another author who had pleaded the same cause, Etienne Bonnot de Condillac, whose empiricist approach at the time of the publication of the *Treatise on systems* (*Traité des systèmes*) in 1749 consisted in admitting only those principles that derived from experience and from carefully gathered data, such as the principle of gravity.[24] For Beausobre, to be sure, abstract principles or 'self-evident truths that can be affirmed of all beings in general and in particular, without exception'[25] are neither

22. Julien Offray de La Mettrie, *Traité de l'âme*, ed. Theo Verbeek, vol.1 (Utrecht, 1988), p.163: 'Il faut faire assidûment les mêmes pas avec la Nature, toujours aidé, comme dit M^{e} la M. du Chattelet, du *Bâton* de l'observation & de l'expérience.' See also La Mettrie's criticism of Leibniz, especially his dismissal of the Principle of sufficient reason as a useless 'principle of system': 'Le principe de la *Raison suffisante* [...] n'est qu'un principe de Système, & fort inutile dans la recherche de la vérité' (*Traité de l'âme*, p.155-56). A similar criticism of the Leibnizian principle is found in Jean D'Alembert's *Preliminary discourse to the Encyclopedia of Diderot*, translated by Richard N. Schwab (Chicago, IL, 1995), p.87: 'His principle of *sufficient reason*, excellent and very true in itself, does not seem very useful to beings so poorly enlightened as we regarding the primary reasons of things.'
23. Maybe Beausobre had personally met La Mettrie, who found refuge in the court of Frederick II, where he lived from 1748 to 1751.
24. Etienne Bonnot de Condillac, *Œuvres philosophiques*, ed. Georges Le Roy, vol.1 (Paris, 1947), p.122.
25. Beausobre, *Dissertations philosophiques*, p.183.

the source of our knowledge nor innate principles. And Beausobre does not seek to deny the empirical foundations of knowledge. What he seeks to make clear is that abstract principles are the only ones that enable us to *demonstrate* anything, to guide us, through reasoning, to the point of either admitting the truth in question or contradicting self-evident principles.[26]

According to him, abstract principles are not to be banned as such because the problem lies not in them but rather in a lack of logic in their application. This lack of logic is apparent in the way they are brought in at the wrong point and in needlessly positing new abstract principles. Knowledge is acquired through the senses, and certainty is obtained by ascending or by having recourse to abstract principles. It is up to the philosopher/scientist to determine where it is appropriate to apply these principles and to use caution in drawing conclusions from them. Experience for its part will with no less certainty lead us into error if we have no principles to guide us and lightly draw a multitude of conclusions that experience seems to suggest to us. According to Beausobre, the search for truth via experiments requires many more precautions than the search for truth via reasoning. Every day, natural philosophers cite experiments and oppose their respective observations to defend contradictory views. In referring *exclusively* to experience, they commit several errors: they suppose that what does not fall under the purview of the senses does not exist; that what precedes an effect is its cause; that if a cause has once produced a certain effect it cannot later produce a contrary effect, and so on. Such prejudices lead experimenters to conduct their work less to find truths than to reinforce their views and hypotheses. Accordingly, there is little hope that experience will prove a sure guide in the search for truth. It is necessary to find a 'golden mean': experience alone, just like reasoning alone, does not suffice to avoid error and find the truth. In this search, we should not be all too ready to distrust our own wits: the impossibility we sometimes find ourselves in to have a distinct idea of certain things must not lead us to incline 'completely' towards Pyrrhonism.[27]

This last affirmation enjoins us to examine whether the search for a golden mean, which does indeed question, at least partially, our trust in experience and abstract principles, is compatible with a moderate form of Pyrrhonism of the kind Beausobre develops in his

26. Beausobre, *Dissertations philosophiques*, p.183-84.

27. Beausobre, *Dissertations philosophiques*, p.205.

1755 *Reasonable Pyrrhonism*.[28] A comparative study of the two texts – the *Dissertations* and *Reasonable Pyrrhonism* – reveals many parallels, more numerous and more significant than might be expected, even if the second text articulates a more pronounced sceptical stance. On the question of abstract principles, we find that *Reasonable Pyrrhonism* does not consider the Principle of sufficient reason as such to be problematic, but only the way it is applied. The difficult thing is always to determine the sufficient reason. When an effect appears in nature, a certain set of principles, taken together, makes it possible to know why it is thus and not otherwise; this knowledge is true, but it always remains a question which principles come together. Error arises from the application of principles, not from the principles themselves – they are not subject to doubt because they make no claims about objects or events.[29] On the question of experience, we find an admission of its necessity: in its absence we would perhaps not know anything at all. Yet once more we also find doubts as to its usefulness as a guide for scientific research: 'But will experience be a light that leads to the truth? What guide could it be when it consumes itself but sheds no light!'[30] Experience is not judged capable of providing us with correct ideas and proving what seems likely to us.

In *Reasonable Pyrrhonism*, Beausobre adds an interesting and, with respect to his earlier work, entirely original remark that renders the status of experience even more complex and problematic. The prejudices of those who have raised us, he affirms, have combined with the imperfections of our senses to produce errors that are destroyed only with great difficulty: in fact, 'nature has given us senses, but it is study and use that have shaped them.' Even more interestingly, the note refers to Condillac's 'excellent' *Treatise on sensations* (*Traité des sensations*) of 1754.[31] Beausobre's remark is an expression of his appreciation of Condillac's mature work, which he considers to be less naive in its conception of experience and better informed in its study of sensation: the senses are themselves constructed by experience, he had claimed; thus experience is not simply intuitive. This appreciation

28. Louis de Beausobre, *Le Pyrrhonisme raisonnable* (Berlin, Etienne de Bordeaux, 1755). A version of this text was published by the author a year earlier (1754), under the title *The Pyrrhonism of the sage* (*Pyrrhonisme du sage*).
29. They do not say anything about how an object or an event realises the conditions formulated by them.
30. Beausobre, *Le Pyrrhonisme raisonnable*, §CXXV, p.178.
31. Beausobre, *Le Pyrrhonisme raisonnable*, §CXXXVII, p.181.

is also understandable because the two authors must have been on rather good terms, as Condillac's rather sportsmanlike reaction to the reproaches addressed to him in Beausobre's *Dissertations*, which he had been charged with reading and approving, suggests: 'The author honours me by contesting what I have said about abstract principles. I might be suspected of affectation if I praised him and of injustice if I criticised him; I have found nothing in his book that would hinder its being printed.'[32]

The mature Condillac is a source for Beausobre, because he allows him further to develop his reflections on the prejudices that can blend in and interfere with experience. Before an object passes into our mind, Beausobre explains, it must take a rather long path, and even in our mind there are prejudices that may disfigure it. And in this same context, he develops a further new aspect of his reflection on the status of experience, dwelling on the structural and prejudicial failings of scientific observations, especially observations made by means of microscopes. Like certain other authors of his time, Beausobre is aware of an important problem in modern science. In particular, he cites the spermatic animal hypothesis that Antonie van Leeuwenhœk had articulated on the basis of microscopical observations, which had subsequently been refuted by Abraham Trembley's discoveries and Pierre-Louis de Maupertuis's objections.[33] Beausobre's conclusion on this point is as laconic as it is emblematic: 'In the experiments we conduct, we usually only see what we seek to find there; we see in them even what cannot be there.'[34]

32. Beausobre, *Dissertations philosophiques*, p.2: 'L'Auteur me fait l'honneur de combattre ce que j'ai dit sur les Principes abstraits. Je serois peut-être soupçonné d'affectation si je le loüois, & d'injustice si le critiquois; je n'ai rien trouvé dans son Ouvrage qui puisse en empêcher l'impression.' There is at least one letter, rather friendly in tone, from Condillac to Beausobre; in this letter, Condillac refers to his *Traité des animaux*, which means the letter must have been written after 1755.
33. Beausobre, *Le Pyrrhonisme raisonnable*, §CXXXVIII, p.182-83: 'Il n'y aura peut-être jamais d'hypothese plus vrai-semblable, que celle des animaux spermatiques, & il n'y en a point qui soit exposée à plus de difficultés: *Leeuwenhœk* n'a cru voir rien de plus certain; cependant la découverte de M. *Trembley* & les objections de Mr. *de Maupertuis* arrêteront tous ceux qui seront équitables.' On the modern debates on animal generation evoked by Beausobre, see Jacques Roger, *Les Sciences de la vie dans la pensée française au XVIIIe siècle*, 3rd edn (Paris, 1993).
34. Beausobre, *Le Pyrrhonisme raisonnable*, §CXLII, p.188.

Does the soul resemble the eyes?

This questioning of the reliability of microscopes resurfaces in the memoirs of the Berlin Academy, and it does so in a passage that is central to understanding Beausobre's thinking not just on the question of experience in general but on that of experience in the specific field of psychology. In the first 'Memoir on the nature and causes of madness' ('Mémoire sur la nature et les causes de la folie'), we find a kind of warning addressed to those seeking the physical causes of madness. This warning reflects Beausobre's polemic with another member of the Academy, the anatomist Johann Friedrich Meckel, who was then studying the phenomenon of madness by means of brain autopsies.[35] Beausobre criticised, on the one hand, a wrong choice of means in physical/anatomical observation and, on the other, an error in the choice of both object and set-up of experiments. As for the choice of object, Beausobre aimed to show that the study of madness must rest not on the observation of the body but on the introspection of the soul; that the imagination is wholly able to account for the follies of the soul such that the soul turns out to be the principle of all its operations; and that it is thus best to remain within and not go beyond the realm of the mind or spirit:

> I think it is important to distinguish between a corpse and a living being, even if the perceptible parts are more or less the same; such is not the case for the almost imperceptible parts, whose form may entirely change and which may even disappear altogether with the cessation of the vital movement. Microscopes may deceive us, and there are many things that can escape the curiosity of the most able experimenter: moreover, an anatomist engaged in such research should also not have neglected the study of sound psychology and should not consider everything settled once movement and matter are supposed.[36]

35. See Johann Friedrich Meckel, 'Recherches anatomico-physiologiques sur les causes de la folie qui viennent du vice des parties internes du corps humain', in *Histoire de l'Académie royale des sciences et belles lettres* (1764; Berlin, Haude and Spener, 1766), p.65-88.
36. Louis de Beausobre, 'Réflexions sur la nature et les causes de la folie, premier mémoire', in *Histoire de l'Académie royale des sciences et belles lettres* (1759; Berlin, Haude and Spener, 1766), p.390-403 (402): 'Je pense qu'il importe de distinguer le cadavre du corps animé, quoique les parties bien sensibles soient à peu près les mêmes: il n'en est pas ainsi des parties presque insensibles, auxquelles la cessation du mouvement vital peut faire changer entierement de forme, et qu'elle peut même faire disparoitre; les microscopes peuvent tromper, & il y

In other words, experience in psychology, which deals with a spiritual being, differs from experience in natural philosophy, which deals with material things.

But how does experience come about in psychology, and how does it relate to reasoning? In the 'Memoir on dreams' ('Mémoire sur les songes') of 1762, Beausobre shows how psychological experience (or introspection) is completed or rectified by applying the Principle of sufficient reason. We already find such a correction of the experience we have of ourselves, of the life of our soul, in *Reasonable Pyrrhonism*: 'Most of our experiences are but judgements about certain facts. How many people say that they are certain, thanks to experience, of not thinking during their sleep, when they should limit themselves to saying that they do not remember having thought during their sleep!'[37] In the 'Memoir on dreams', Beausobre sets out to prove that, even if gaps in our memory caused by obscure ideas make it seem otherwise, dreams belong to the soul's chain of representations. The Principle of sufficient reason plays a decisive role here, as it had in Wolff's treatment of the question. Wolff had argued that, when one considers dreams subjectively, with respect to this and that 'phantasm' coexisting in the soul, or phantasms succeeding one another, the rules of the imagination can account for and explain both the law of coexistence and the mechanism of succession: the mutations that affect the dreaming soul are not deprived of a sufficient reason. The soul thus follows a constant law, and, whenever it imagines anything, nothing happens without a sufficient reason. If, however, one considers dreams objectively, with regard to phantasms that represent things as being present outside of us, the mutations of these things take place without a sufficient reason.[38] Beausobre was thus entirely justified in having recourse to the Principle of sufficient reason, even if this principle had traditionally been used mainly to mark the difference, from the objective point of view, between dreams and reality. In dreams, he notes:

> An obscure idea becomes clear; the transition from obscurity to clarity appears abrupt to us because we do not perfectly see which

a bien des choses qui peuvent échaper à la curiosité du plus habile: il faudrait aussi qu'un Anatomiste occupé de cette recherche, n'eût point négligé l'étude d'une saine Psychologie, & ne crût pas tout dit lorsqu'on suppose du mouvement & de la matiere.'

37. Beausobre, *Le Pyrrhonisme raisonnable*, §CXLIII, p.189.
38. Wolff, *Psychologia empirica*, §128, p.82-83.

> association of ideas, partly clear, partly obscure, has reproduced this new idea. Everything in nature is connected, but not in such a way that we never miss the connection. Since there is nothing without a sufficient reason, however strange certain events may seem to us, they are just as natural as those that seem to us to conform the most to the common laws: they differ from these latter only through our way of seeing: the reasons that explain what they are and how they are what they are, these reasons are not as known to us as those that explain the events that have nothing striking about them for us: in the latter case, we are more enlightened, in the former more ignorant; that is the difference.[39]

If what happens in dreams is not without a sufficient reason, it is clear why these phenomena, too, depend on the common laws of the imagination and why it would be wrong to consider them unrelated to the other representations of the soul or to associate them with causes other than the laws that govern the activity of the soul itself.

Yet there is another aspect worth considering that characterises experience as well as the cooperation between reasoning and experience in the psychological domain. In Beausobre's 1768 'Reflections on the nature and necessity of obscure ideas' ('Réflexions sur la nature et la nécessité des idées obscures'), which can be seen as outlining a kind of general theory, or at least a summary of the fundamental ideas that form the basis of his other contributions, we read:

> The soul is a force representative of the universe: all the ideas it can have are currently within it, but, since it is a finite force, all these ideas cannot be there to an equal degree of clarity; indeed, there are few that succeed in being clearly seen, and there is always one that [is more clearly seen] than all the others: the light of one idea eclipses all

39. Louis de Beausobre, 'Réflexions philosophiques sur les songes, premier mémoire', in *Histoire de l'Académie royale des sciences et belles lettres* (1762; Berlin, Haude and Spener, 1769), p.429-40 (435): 'Une idée obscure devient claire; le passage de l'obscurité à la clarté nous paroit brusque, parce que nous ne voïons pas parfaitement par quelle association d'idées, en partie claires en partie obscures, cette nouvelle idée a été reproduite. Tout est lié dans la nature, mais tout ne l'est pas de façon que la liaison ne nous échappe jamais. Comme il n'y a rien sans raison suffisante, quelqu'étranges que nous paroissent certains événemens, ils sont tout aussi naturels que ceux qui nous paraissent les plus conformes aux loix ordinaires: ils ne diffèrent de ces derniers que par notre maniere de voir: les raisons qui expliquent ce qu'ils sont & comment ils le sont, ne nous sont pas aussi connues que celles qui expliquent ces événemens qui n'ont rien de frappant pour nous: ici nous sommes plus éclairés, là plus ignorans; voilà la différence.'

the others. *Our soul resembles our eyes, we cannot see everything, and what we cannot see we must suppose.*[40]

Why does Beausobre, in this last sentence, affirm that all we can do is formulate hypotheses and content ourselves with conjectures? We have seen that as early as 1753, in the *Dissertations*, he had used a very similar expression ('we do not see everything') in talking about the limits of experience in natural philosophy. Yet there, abstract principles were able to make up for the failures of experience: they allowed for demonstration and for obtaining certainty; one was not obliged to limit oneself to supposing.[41]

How can we account for the difference between the ways the status of attainable knowledge is defined in the writings on natural philosophy and in the writings on psychology? There are at least three possible interpretations:

(1) The difference is not really a difference at all. The eye, in natural philosophy, does not see everything, just as the soul, in psychology, does not see all: in both cases, the application of the Principle of sufficient reason can act as a remedy. The originality of the 'Memoirs' over against the *Dissertations* would simply consist in combining introspection (experience) and the Principle of sufficient reason (rationality) to show the continuity between ordinary and (apparently) extraordinary phenomena.

(2) The difference is a result of the sceptical positions Beausobre defends in his mature philosophy. In *Reasonable Pyrrhonism*, he had in fact asserted that the careful search for the truth will not procure certainty. We cannot flatter ourselves with knowing anything with certainty; we thus have to renounce certainty and settle for 'seeing confusedly through a curtain and conjecturing what might be hidden

40. Louis de Beausobre, 'Réflexions sur la nature et la nécessité des idées obscures', in *Histoire de l'Académie royale des sciences et belles lettres* (1768; Berlin, Haude and Spener, 1770), p.411-18 (412), my emphasis: 'L'ame est une force représentative de l'Univers: toutes les idées qu'elle peut avoir sont actuellement en elle; mais comme elle est une source finie, toutes ces idées ne peuvent pas y être avec un degré égal de clarté: il y en a même peu qui parviennent à être clairement apperçues, & il y en a toujours une qui l'est plus que toutes les autres; la lumière de l'une éclipse les autres. Notre ame ressemble à nos yeux, on ne peut pas tout voir, & ce qu'on ne voit pas on doit le soupçonner.'

41. Provision is made for the fact that natural philosophy does not enjoy the same degree of certainty as mathematics, which is the only domain in which it is possible to speak of demonstration in the strict sense. On this point, see Beausobre, *Dissertations philosophiques*, p.223.

there'. Conjectures thus take the place of truth; in the absence of what is certain, we have to settle for what is probable.[42] The short psychological texts written after *Reasonable Pyrrhonism* (whereas the *Dissertations* were written prior to it) thus might convey the author's declining confidence in the powers of the mind, prompted by his work on scepticism.

(3) The difference in question could also reside not in the chronology of the works but simply in the heterogeneity of the two disciplines, in the specificity of psychology and its empirical branch over against the experimental part of natural philosophy. The search for a golden mean in psychology would then be distinct from that in the field of natural philosophy. Wolff himself had already nuanced the perfect symmetry of psychology and natural philosophy when he pointed to a difference in what constituted experience in the two disciplines: the *ars observandi* draws truths from observations and is employed by natural philosophers and psychologists alike; the *ars experimentandi*, in turn, draws truths from experiments and is employed only by physicists, at least for the moment.[43] Introspection, accordingly, is only an art of observation, one in which the coincidence of observer and object observed implies a lack of experimental verification; when compared to natural philosophy, it suffers from a lack of certainty.

Unexpected convergences

It is not easy to give preference to one of these interpretations in particular, at least at first glance. If we take into consideration the implications and the heuristic scope of each of them, the task will nevertheless become easier. The first and second hypotheses tell us something about Beausobre himself, that is, about the continuity or discontinuity in his own career, while the third enlarges our views by making possible a comparison, which places Beausobre in the context of the Berlin Academy. More specifically, the second hypothesis is an argument for discontinuity, according to which the publication of the *Reasonable Pyrrhonism* would have played a major role in reorienting Beausobre's thinking after 1755. But my inclination rather tends to view the importance of this work as a turning point, so I would be inclined to turn to the remaining hypothesis, which seems to be free from such a defect. On the one hand, the first interpretation drives

42. Beausobre, *Le Pyrrhonisme raisonnable*, §CLXII, p.212.
43. Wolff, *Psychologia empirica*, §458-59, p.357-58.

us to interpret the analogy that, in his different texts, Beausobre establishes between the eye and the soul as coherent with what John Locke says in his *Essay*, in the 'Epistle to the reader', about boundaries that intellection shares with vision: 'For the Understanding, like the Eye, judging of Objects only by its own Sight, cannot but be pleased with what it discovers, having less regret for it has scaped it, because it is unknown.'[44] We would thus be dealing with a kind of moderate gnoseological scepticism, which would cross the whole work of Beausobre by giving it a certain degree of homogeneity, at least from a methodological point of view. On the other hand, the third hypothesis invites us to reformulate, or rather to complete, Beausobre's analogy by having recourse to an ancient *topos* and affirm that 'like the eye, the soul, though it does not see itself, yet discerns other things.'[45] Locke himself had already incorporated this additional element in the *Introduction* to the first book of his *Essay*: 'The Understanding, like the Eye, whilst it makes us see, and perceive all other Things, takes no notice of it self: And it requires Art and Pains to set it at a distance, and make it its own Object.'[46] It is precisely the same version of the analogy between the soul and the eye that we find in the texts of another member of the Academy: Jean Bernard Merian (1723-1807), who is undoubtedly one of the most interesting figures among the members of the class of speculative philosophy.[47] Merian devoted his memoirs to several disciplines, including history of philosophy and the different parts of metaphysics (ontology, psychology and morals). His remarks on the epistemological status of psychology deserve to be recalled here as they can shed new light on Beausobre's opinions.

Merian, too, did notice the basic problem of introspection, that is the coincidence between the subject who observes and the object to

44. John Locke, *An Essay concerning human understanding*, ed. Peter H. Nidditch (Oxford, 1975), p.6.
45. Cicero, *Tusculan disputations*, translated by J. E. King (Cambridge, MA, 1945), I.xxvii.67, p.78-79.
46. Locke, *Essay*, I.i.1, p.43.
47. On Merian and the sciences of the soul, see Bernard Bærtschi, 'La conception de la conscience développée par Mérian', in *Schweizer im Berlin des 18. Jahrhunderts*, ed. Martin Fontius and Helmut Holzhey (Berlin, 1996), p.217-30, and Bernard Bærtschi, *Conscience et réalité: études sur la philosophie française au XVIIIe siècle* (Geneva, 2005), while on his eclecticism, see notably Daniel Dumouchel, 'Hommes de nulle secte: éclectisme et refus des systèmes chez Jean Bernard Mérian', *Dialogue* 57:4 (2018), p.745-65, and Tinca Prunea-Bretonnet's chapter in this volume. See also Stefanie Buchenau, '"L'œil ne se voit pas voir": Sulzer sur la contemplation et le sentiment de soi', *Philosophiques* 42:1 (2015), p.73-88.

be observed, which affects psychology. In order to express this idea, Merian first evokes the image of Janus, as it is attested by a striking passage taken from his 1749 memoir 'On the apperception of one's own existence' ('Sur l'apperception de sa propre existence'):

> The following proposition: *I perceive my own apperception*, is contradictory; for it would be necessary for me to be at the same time myself and another, and that I might, as it were, perceive through myself, my own apperception – roughly, in the same way as we look at objects through binoculars – or it would be necessary, like the god Janus, to look forward, and backwards, that is to perceive, and to reflect at the same moment. If that were the case, the perceiving being could not only be doubled, but also for the same reason multiplied infinitely: absurdities which are too palpable to deserve to be considered.[48]

The analogy between the soul and the eye resurfaces a few years later, in Merian's 'Discourse on metaphysics' ('Discours sur la métaphysique') of 1765:

> 'Nothing is better known to us than the soul', says Leibniz, 'because the soul is intimate to us, that is, intimate to itself.' But is it not necessary for the mind, as for the eye, to keep its distance in order to distinctly perceive objects? What pain does it feel in turning back to itself? These kinds of contemplation are so difficult and so uncertain, because they concern the very subject which contemplates, because we are, at one and the same time, the observers and the thing to be observed. Psychology does not operate by artificial projections.[49]

48. Jean Bernard Merian, 'Mémoire sur l'apperception de sa propre existence', in *Histoire de l'Académie royale des sciences et belles lettres* (1749; Berlin, Haude and Spener, 1751), p.416-41 (417): 'La proposition: *je m'apperçois apercevoir*, est contradictoire; car il faudroit, pour que cela fut, que je fusse en même tems Moi-même & un autre, et que je pusse, pour ainsi dire, appercevoir à travers de moi-même appercevant; à peu près comme on regarde les objets au travers de lunettes d'aproche; ou il faudroit, comme le Dieu Janus, regarder en avant, & en arrière, c'est-à-dire, appercevoir, et réfléchir dans le même instant. Si cela étoit, non seulement l'être appercevant pourroit se doubler, mais encore par la même raison se multiplier à l'infini: absurdité trop palpable pour mériter qu'on s'y arrête.'
49. Jean Bernard Merian, 'Discours sur la métaphysique', in *Histoire de l'Académie royale des sciences et belles lettres* (1765; Berlin, Haude and Spener, 1767), p.450-74 (456): '"Rien ne nous est mieux connu que l'ame," dit Leibnitz, "parce qu'elle nous est intime, c'est à dire intime à elle-même." Mais ne faut-il point à l'esprit, comme à l'œil, une certaine distance pour appercevoir distinctement ? Quelle

In this memoir, Merian insists on the specificity of metaphysics (of which psychology is a part), by distinguishing it not only from mathematics and geometry,[50] but also from natural philosophy. Moreover, he gives a particular emphasis to the problems connected with the study of the mind, which are of little relevance to the study of bodies.[51] According to him, the objects of metaphysics have a subtlety, a delicacy which can even embarrass people who are exercised in meditation. In fact, sometimes they vanish before they are fixed, sometimes they darken one another, often there is only one moment favourable to seize them and it is almost never possible to repeat the observation of a phenomenon: when we return to it, the subject has undergone alterations. So it seems difficult to rigorously observe ourselves, as we should be placed outside ourselves and be able to contemplate past events. This is why, in metaphysics, meditation often bears no fruit, while in natural philosophy we can always hope to succeed by stubborn work. Bodies lend themselves to our senses when we want to observe them, because we are dealing with permanent things or at least with phenomena which recur frequently; the differences and the relations between the objects of observation are easily marked, and the natural philosopher, like the chemist, the anatomist or the astronomer, can force nature to give them an answer and can count on many auxiliary resources.

In his 'Discourse on metaphysics', Merian also establishes a classification of the sciences and establishes a number of relationships

peine ne sent-il pas à se replier sur lui-même? Ces sortes de contemplations ne sont si difficiles & si incertaines que parce qu'elles se font sur le sujet même qui contemple, parce que nous sommes, tout à la fois, les observateurs & la chose à observer. La Psychologie n'opere pas par des projections artificielles.'

50. In the debate on the use of the mathematical method in philosophy at the Berlin Academy, which culminated in the *Preisfrage* for 1763, Merian aligns on the side of those who support the irreducibility of metaphysics to mathematics and geometry. He is a partisan of the anti-Wolffian party, which follows the suggestions given by Maupertuis in his *Examination of the proofs of the existence of God* (1756). According to Maupertuis, in fact, the mathematical or geometrical method cannot be transposed into the field of metaphysics. On this debate see notably Jean-Paul Paccioni, 'Présentation', *Astérion* 9 (2011), special issue: *La Preisfrage de 1763 de l'Académie de Berlin: certitude mathématique, certitude métaphysique, certitude morale*, ed. Jean-Paul Paccioni, https://doi.org/10.4000/asterion.2090, and Tinca Prunea-Bretonnet, 'La méthode philosophique en question: l'Académie de Berlin et le concours pour l'année 1763', *Philosophiques* 42:1 (2015), p.107-30.

51. Merian, 'Discours sur la métaphysique', p.456-57.

among them. In addition to being a part of metaphysics (with cosmology and natural theology), psychology shares something with natural philosophy.[52] We could therefore consider metaphysics as 'the physics of immaterial beings'. Undoubtedly, Merian establishes this parallelism between metaphysics and natural philosophy in the wake of Locke and Lockeanism. In fact, his statements are very close to the ones of Jean D'Alembert in the *Preliminary discourse* of the *Encyclopedia*, according to which Locke had the great merit of reducing metaphysics to the experimental natural philosophy of the soul.[53] But such a parallelism is possible and legitimate, from the perspective of Merian himself, because the status of natural philosophy is not exactly equal to that of mathematics and geometry. In this connection, it should be remembered that, according to Pierre-Louis Moreau de Maupertuis (whom Merian seems to follow), there is a science, so to speak, intermediate between mathematics and metaphysics, namely, dynamics, which treats of physical objects and participates only in part in the evidence and certainty of arithmetic and geometry.[54] How did Merian develop this last connection between certain parts of metaphysics (including psychology), on the one hand, and natural philosophy, on the other? From his point of view, in natural philosophy we sometimes prove according to hypotheses, but these hypotheses, although they reach the highest possible degree of probability, are not susceptible of demonstration. The three transcendent sciences corresponding to natural philosophy are hampered by exactly the same problem: they are not rigorously demonstrable, because the nature of their proper objects resist demonstration. The two domains thus compared share certain limits and have a common basis; indeed, observation and experience are essential for the science of bodies as well as for the study of immaterial beings.[55]

52. Merian, 'Discours sur la métaphysique', p.459.
53. D'Alembert, *Preliminary discourse*, p.83-84.
54. Pierre-Louis Moreau de Maupertuis, 'Examen philosophique de la preuve de l'existence de Dieu employée dans l'*Essai de cosmologie*', in *Histoire de l'Académie royale des sciences et belles lettres* (1756; Berlin, Haude and Spener, 1758), p.389-424 (399-400).
55. Merian, 'Discours sur la métaphysique', p.456. We have already noted (see n.41) that, according to Beausobre, natural philosophy does not share the same degree of certainty as mathematics. This stance dates from a few years before the methodological debate broke out between Wolffians and anti-Wolffians (see n.50).

Conclusions

In Merian's memoirs, the methodological reflections on the use of rational tools (notably hypotheses), which should work alongside experience in natural philosophy as well as in psychology, are quite developed and lead to some original conclusions. In Beausobre's texts, the methodological reflection is perhaps less formalised but it is constant over time. So, in spite of their opposite affiliation (Merian was an anti-Wolffian and a follower of Maupertuis, while Beausobre explicitly adheres to Formey's Wolffianism), both these authors were interested in the relation between reasoning and experience and came to quite similar conclusions concerning the status of natural philosophy and psychology. Of course, Beausobre does not underestimate the importance of rational tools such as the Principle of sufficient reason articulated by Leibniz and Wolff. Yet at the same time, he shares with other Enlightenment authors the conviction that there is no way of circumventing experience. I have referred above all to his ambivalent attitude concerning Condillac: while he supports Condillac's radicalisation of Lockean empiricism in the *Treatise on sensations* of 1754, he contests the absolute refusal to acknowledge any abstract principles attested by the *Treatise on systems* of 1749. Even more interesting in this context is Beausobre's critical view of the questions at issue in modern scientific debates. With regard to research in physics, microscopical observations in the life sciences, and medical research on the corporeal correlates of the intellectual faculties, he wonders if experience really is, in all instances, the 'light of truth', or whether it might not at times be subject to error and biased by prejudice in the same way reason is. Finally, I have shown that, in the context of the Berlin Academy, Beausobre was not the only author who noticed a tension between proximity and distinction while comparing the domains of natural philosophy and psychology. What brings these fields closer together is, without any doubt, the necessity to search for a golden mean, for neither natural philosophy nor psychology enjoys the privileges of geometry and mathematics. What would instead separate the science of bodies from the science of the soul is probably a more rigorous and specific definition of the golden mean itself, a definition which remains, however, in both cases, a goal we are yet to reach.

Maupertuis, Euler and the Leibnizian metaphysics behind the Principle of least action

Ansgar Lyssy

Introduction

It is no exaggeration to say that Pierre-Louis Moreau de Maupertuis (1698-1759) and Leonhard Euler (1707-1783) dominated the Berlin Academy in the 1740s and 1750s. This is by no means trivial, for scientific academies, such as that in Berlin, and their journals shaped the content, method and outreach of Enlightenment science and, thereby, of the Enlightenment itself, which often went hand in hand with the development of modern science. The Berlin Academy was famously founded by G. W. Leibniz and started out following his plans and ideas about the nature of science and the nature of the Republic of Letters. But Leibnizianism was the subject of an intense debate that shaped both the development of modern science and the academies themselves. Both Maupertuis and Euler had an ambivalent relationship with Leibniz's philosophy and science. Thus, it is reasonable to assume that their transformation of Leibnizian ideas, which proved highly influential for the development of modern science, would also shape the scope and outreach of research done in the Berlin Academy.

In this chapter, I will point out the ambivalent and tense relationship Maupertuis and Euler had to their predecessor, especially concerning Leibniz's grounding of physics in metaphysics. Consequently, this chapter has two intersecting goals: first, it attempts to flesh out some aspects of the reception of Leibnizian thought in the context of Enlightenment physics, more precisely, in the deduction of the Principle of least action (henceforth PLA).[1]

1. This chapter is only concerned with the metaphysical background of the emergence of the PLA within a rather specific time frame. For the mathematical aspects, see Marco Panza, 'De la nature épargnante aux forces généreuses: le

Second, it also highlights a specific approach towards the intersection of physics and metaphysics that was championed by the Berlin Academy, which would not have been well received in Paris or London. To that end, I will briefly examine some aspects of the strategies of argumentation, metaphysical presumptions and self-ascribed methodologies (often quite different from their actual methodology) employed by Maupertuis and Euler in their respective deductions of the PLA.

One of the innovations of the Berlin Academy was the introduction of a class of speculative philosophy, mostly of epistemological and metaphysical speculations, which was notably absent from its main rival at that time, the French Academy. Nor was any equivalent to be found at the Royal Society of London. The Academy comprised four different classes, covering the division of the sciences and arts: *experimental philosophy*, which relates to all knowledge gained from (systematic) experience; *mathematics*; *speculative philosophy*; and *belles-lettres*, which is concerned with all issues concerning language, literature and the arts. Consequently, the Academy's annual publication, entitled *Histoire de l'Académie royale des sciences et belles lettres* until 1769, not only printed the usual selection of papers in optics, mechanics, astronomy, botany, medicine and so on, but also contained papers on, to name just a few examples from the earlier volumes of the *Histoire*, Anaxagoras,[2] Spinoza,[3] the existence of God[4] and the moral usefulness of science.[5]

---

principe de moindre action entre mathématiques et métaphysique – Maupertuis et Euler, 1740-1751', *Revue d'histoire des sciences* 48:4 (1995), p.435-520; Helmut Pulte, *Das Prinzip der kleinsten Wirkung und die Kraftkonzeptionen der rationalen Mechanik* (Stuttgart, 1989); Alberto Rojo and Anthony Bloch, *The Principle of least action, history and physics* (Cambridge, 2018); and Wolfgang Yourgrau and Stanley Mandelstam, *Variational principles in dynamics and quantum theory*, 2nd edn (New York, 1960). For an interpretation of the PLA within the context of modern science, see, for example, Cornelius Lanczos, *The Variational principles of mechanics* (Toronto, 1949); Jennifer Coopersmith, *The Lazy universe: an introduction to the Principle of least action* (Oxford, 2017).

2. Johann Philipp Heinius, 'Vie d'Anaxagore, philosophe très célèbre dans l'école ionique', in *Histoire de l'Académie royale des sciences et belles lettres* (1752; Berlin, Haude and Spener, 1754), p.313-49.
3. Philipp Joseph von Jariges, 'Examen du spinozisme et des objections de Mr. Bayle contre ce système', in *Histoire de l'Académie royale des sciences et des belles lettres de Berlin* (1745; Berlin, Haude, 1746), p.121-42.
4. Jean Henri Samuel Formey, 'Les preuves de l'existence de Dieu, ramenées aux notions communes', in *Histoire de l'Académie royale des sciences et belles lettres* (1747; Berlin, Haude and Spener, 1749), p.341-64, and 'Examen de la preuve qu'on tire

see opposite for footnote 5

This specific outline of the sciences was considered peculiar and thus in need of justification. The unnamed editor of the first volume of the *Histoire de l'Académie royale des sciences et des belles lettres de Berlin* published in 1746, presumably Jean Henri Samuel Formey, acknowledges that the existence of a class of speculative philosophy requires special justification. Metaphysics, they say, is 'the mother of the other sciences, the theory that provides the most general principles, the source of evidence and the foundation of the certainty of our knowledge'.[6] This should not be confused with the metaphysics of the scholastics, which hardly produced anything other than 'brambles and thorns' (*des ronces et des épines*) and which is the reason why metaphysics has such a bad reputation. But metaphysics need not be that way. In modern times, metaphysics has had a guiding and foundational function, as it provides the principles of scientific enquiry and thought. This brief remark can be understood as fulfilling a double function. First, it detaches metaphysics from theology, as it breaks off with the entire scholastic tradition, which was still important to the theology of that time. Metaphysics needs some relation to science or other, more practical forms of knowledge, lest it become a vain exercise in thinking, a 'glass bead game' of sorts. Second, the editor evokes the idea of an axiomatic and deductive physics, which starts out with and is grounded in metaphysical principles. These principles can both ground all knowledge in apodictic certainty and guide the practice of scientific enquiry. This broadly formulated approach derives from Aristotle and Bacon, but was heavily supported and shaped by Descartes and Leibniz. While this approach was largely common ground at that time, the details were important and contested: the separation of natural and experimental philosophy opens up new

des fins de la nature, pour établir l'existence de Dieu', in *Histoire de l'Académie royale des sciences et belles lettres* (1747; Berlin, Haude and Spener, 1749), p.365-84. See also Pierre-Louis Moreau de Maupertuis, 'Examen philosophique de la preuve de l'existence de Dieu employée dans l'*Essai de cosmologie*', in *Histoire de l'Académie royale des sciences et belles lettres* (1756; Berlin, Haude and Spener, 1758), p.389-424.

5. Jean Henri Samuel Formey, 'Examen philosophique de la liaison réelle qu'il y a entre les sciences et les moeurs', in *Histoire de l'Académie royale des sciences et belles lettres* (1753; Berlin, Haude and Spener, 1755), p.397-416.

6. 'Préface', in *Histoire de l'Académie royale des sciences et des belles lettres de Berlin* (1745; Berlin, Haude, 1746), sig.)( )(2*r*; see Mary Terrall, *The Man who flattened the earth: Maupertuis and the sciences in the Enlightenment* (Chicago, IL, 2002), p.265. All translations are my own unless otherwise stated.

pathways of evaluating and criticising metaphysics in regard to its evidence and its function as a foundation of certainty. The role of mathematics also needs to be reconceptualised – is it the basis for some enquiries or a mere tool? As we will see in the following sections, Maupertuis and Euler come to quite different conclusions.

Maupertuis makes a strong programmatic statement in his address 'On the duties of the academician'.[7] Speculative philosophy deals with abstract matters, or 'considers those objects which no longer have any bodily property'.[8] With the qualifier 'no longer' ('ne … plus') he indicates a process of abstraction, by which we infer abstract objects from our ideas by disregarding all sensible or physical properties, indicating a commitment to experimental philosophy as a general approach. What qualifies as the subject of speculative philosophy is the highest being, the mind and all that belongs to it, as well as the nature of bodies, in so far as it must be considered to be different from their phenomena. But as these topics are not grounded in sensory experience, they are prone to uncertainty and error and the philosopher thus needs to distinguish between the different degrees of certainty, probability and sceptical doubt that can and must be assigned to the different forms of knowledge, depending on how they are derived. While Maupertuis here displays his empiricist influences, namely in asserting that certainty can only be found in sensory knowledge, we will later see that Euler instead tried to avoid founding any theory on experience. The question of the epistemological status of metaphysical principles will be dealt with later. Maupertuis finishes his address with a brief remark that not one single discipline but all of them together should lead to the study of religion and the conversion of the infidels,[9] including speculative philosophy, which leads us to God. However, it is clear that philosophy here has long parted ways from theology; it has already established itself to be more than the

7. Pierre-Louis Moreau de Maupertuis, 'Des devoirs de l'académicien', in *Histoire de l'Académie royale des sciences et belles lettres, depuis son origine jusqu'à présent* (Berlin, Haude and Spener, 1750), p.115-25. See the appendix to this volume for an English translation of Maupertuis's address.

8. 'La Philosophie Spéculative considère des objets qui n'ont plus aucune propriété des corps'; Maupertuis, 'Des devoirs de l'académicien', p.120.

9. Maupertuis, 'Des devoirs de l'académicien', p.124: 'Le premier Réglement de la Societé Royale portoit, qu'une de ses Classes devoit s'appliquer à *l'Etude de la Religion & à la conversion des Infideles* […]. Notre Réglement moderne ne charge aucune Classe en particulier de cette occupation: mais ne peut-on pas dire que toutes y concourent?'

mere 'handmaiden' of theology and has become the leading science or *Leitwissenschaft*.[10] I will come back to this later.

But what does this mean for the relation between philosophy and natural philosophy – in the present case, physics? To make this point a bit clearer, I want very briefly to sketch some aspects of the fundamental opposition in scientific and philosophical methodology of the time, namely the opposition between Leibnizianism and Newtonianism. The former had a strong foothold in Germany, while the latter was a dominant feature in French and English thought of the period.[11] A comprehensive and nuanced comparison is, of course, not possible within the course of this chapter, but at least I can highlight some individual aspects that may serve as a framework for the following discussion.[12] It should be clear that this framework is reduced to a selection of topics that will be of importance later. As I am only concerned with the PLA, I will also have to ignore the contributions Newton made to 'save' optical and astronomical phenomena, as well as his reconciliatory account of experimentation and mathematics. I will also ignore the debate about the origins of the calculus.[13]

10. For a survey of the emancipation of philosophy in the eighteenth century, see Hans-Erich Bödeker, 'Von der "Magd der Theologie" zur "Leitwissenschaft": Vorüberlegungen zu einer Geschichte der Philosophie des 18. Jahrhunderts', *Das 18. Jahrhundert* 14:1 (1990), p.19-57.
11. There is now a great amount of literature available for the development of Newtonianism in Europe, especially for Newtonianism as a catalyst for the Enlightenment. For example, see Ronald S. Calinger, 'The Newtonian–Wolffian controversy: 1740-1759', *Journal of the history of ideas* 30:3 (1969), p.319-30; John L. Greenberg, *The Problem of the earth's shape from Newton to Clairaut* (Cambridge, 2010); Margaret C. Jacob, 'Newtonianism and the origins of the Enlightenment: a reassessment', *Eighteenth-century studies* 11:1 (1977), p.1-25; Pulte, *Das Prinzip*; J. B. Shank, *The Newton wars and the beginning of the French Enlightenment* (Chicago, IL, 2008).
12. I will speak of both positions as 'isms': Leibniz's contemporaries and followers had a rather distorted view of his philosophy, as many crucial texts were published only much later; and many of Newton's followers rather chose to focus on his *Principia*, ignoring many of his physico-theological ideas that grounded the mathematics of the *Principia* in an ontology that would not fit well into the framework of Enlightenment materialism. Thus, the picture I present is clearly a simplified one; and 'Leibnizianism' designates a philosophy that overlaps with, but is not identical to, what we today know about Leibniz's thought.
13. There is a vast body of research on Newton's impact on the philosophy of science, too vast to give an overview here. To name just a few publications of interest that deal with these topics: I. Bernard Cohen, *Introduction to Newton's*

One crucial difference between Leibnizianism and Newtonianism concerns the theory of forces. The very name 'dynamics' originates from Leibniz, who uses it to distinguish the science of forces from the purely geometrical approach to motion, which would rather be called 'phoronomia'. It is an enquiry into the nature of forces conceived as one 'level' of reality, even though Leibniz repeatedly makes it clear that the forces apparent on the phenomenal level are of a derivative nature, originating from substances, which are the most basic level of reality. It is an enquiry into proper causes in nature, while mechanics can only display the *how*, not the *why*. Forces are conserved in any event; thus the initial state can always be restored, and as such we assume some entities underlie the interaction in such a harmonious way that all causes and effects are equipollent. Substances are essentially active in themselves and, consequently, all change originates from them via forces. Any apparent passive force is just a reduction of an original activity into a diminished action.

Newton, by contrast, conceives of his physics as a demonstrative science of motion called *mechanica rationalis* – a mathematised natural philosophy without too much metaphysical ballast, clearly not a dynamics in the Leibnizian sense. For Newton, natural philosophy needs to investigate the mathematical formulas to describe forces by means of the phenomenon of motion. Metaphysical speculations about the nature of forces themselves are both unnecessary and unwanted within the realm of physics. Consequently, he does not care so much about the ontological status of these forces, whether they are self-sufficient entities or dependent on others, whether they relate to simple concepts or are conceptually derived from something, and other questions that are key for Leibnizianism. Rather, Newton is interested in changes of state (i.e. how a body transits from motion to rest or vice versa). He distinguishes between *vis insita* (power of resistance to change) and a directionless *vis impressa* (instantaneous and ephemeral action capable of changing the state of relative motion). Forces, conceived as transient external causes, are needed to explain changes of state. Furthermore, while Newton's mechanics is designed to hold for rigid bodies that should not be considered inherently active, Leibniz denies that absolutely hard bodies exist,

---

'Principia' (Cambridge, 1971); Andrew Janiak, *Newton as philosopher* (Cambridge, 2008); Ernan McMullin, 'The impact of Newton's *Principia* on the philosophy of science', *Philosophy of science* 38:3 (2001), p.327-38; Richard Westfall, *The Construction of modern science* (Cambridge, 1971).

and he develops a dynamics modelled around the collision of elastic bodies, which all have an inherent tendency to motion. For Leibniz, causation is fundamentally immanent, and transient causation is merely a structure within the phenomenal world. Newton on the contrary believes only in transient causes. This also relates to the fundamental outline of physics: the law of inertia and the composition of motion, for example, are two of the cornerstones of Newton's mechanics, while Leibniz is less interested in them.

The big advantage Newton's theory has over Leibniz's theory lies in its lack of metaphysical ballast, as it only depends on the truth of its mathematical formulas and their applicability to complex phenomena. But while Newtonians can mock Leibnizians for their metaphysical and speculative substantial activity, they have to accept an unexplainable and unintelligible force instead, namely gravity or action at a distance. The problem that gave rise to the *vis viva* controversy seems to lie in this conflict between the mathematical formulas and the metaphysical assumptions involved: Leibniz's formula $f = mv^2$ is essentially correct, and the French natural philosophers, keen on keeping metaphysics out of science, thus had to take care that mv^2 (or *vis viva*, as the corresponding force was called in regard to its metaphysical status) was shown to be either of little importance in physics, or a derivative concept, or somehow not metaphysical.

That being said, another of the major differences in the debate between Leibnizianism and Newtonianism concerned how to deal with explanations and the question of 'grounding'. Leibniz had insisted that physics needed to be grounded in metaphysics, for otherwise the central notions of physics, especially that of force, body, space and time, would remain unintelligible. They also needed to be grounded in more primitive notions for justification. Any causal explanation would, lest it remained inconclusive, always end up referring to two things: first, one needed to at least imply superior principles that explain why causal principles actually can serve as a valid explanation, such as the Principle of sufficient reason; and second, to achieve a full explanation of any event, one would need to trace any causal explanation back to the divine decision to create this world and no other, as God is the cause of all causes (*causa causarum*). Thus, all scientific explanations would necessarily rely on metaphysics and even theology – something that most Newtonians, even if not Newton himself, would rather avoid.

In the following, I will dissect some of the arguments proposed for the PLA, which is the most famous and influential single idea developed and discussed in the writings of the Berlin Academy. Then I will give a brief discussion as to how the method and theory proposed by Maupertuis and Euler fit with the above-mentioned Leibniz–Newton opposition, and how far this can reasonably be seen as representative of the Berlin Academy.

Maupertuis

The French mathematician and philosopher Pierre-Louis Moreau de Maupertuis introduced the PLA in a paper called 'Accord de différentes loix de la nature qui avoient jusqu'ici paru incompatibles' ('Accord between different laws of nature which at first seemed incompatible', 1744) and further developed his own account in 'Les loix du mouvement et du repos déduites d'un principe métaphysique' ('The laws of motion and rest deduced from a metaphysical principle', 1746), which in turn is a reaction to Euler's mathematical take on it in his treatise *Methodus inveniendi lineas curvas maximi minimive proprietate gaudentes* (*A Method for finding curved lines enjoying properties of maximum or minimum*, 1744). He was profoundly inspired by Newton's works, but also by Leibniz's metaphysical idea of an economy of creation, both of which play an important role in Maupertuis's deduction of the PLA. The abstract concept of action was known before, mostly as a vague mechanical concept in the Newtonian axiom *actio* = *reactio*, but also in Leibniz, who had introduced an *actio motrix* as the expression of force over time.[14] Maupertuis was the first to give it a more precise mathematical expression and to formulate it as a general principle of nature.

Maupertuis deduces the PLA in the context of optics, and then extends it into the domain of mechanics. He arrives at the PLA by criticising and expanding on Fermat's optical theorem that light always takes the shortest path, since Fermat had failed to provide a reason why light should take either the shortest or the fastest route. Maupertuis writes:

> After meditating deeply on this topic, I have thought that light, when it passes from one medium to another, already abandoning the

14. The earliest version of the concept of least action probably originates in a letter from Leibniz, which has been lost and whose authenticity was intensely debated even at the time of the Academy. For a comprehensive account of this debate, see Ursula Goldenbaum, *Ein gefälschter Leibnizbrief? Plädoyer für seine Authentizität* (Hanover, 2016).

shortest path, which is that of the straight line, may as well not follow the path of least time. Indeed, what preference should there be for time over space? Since light cannot travel both by the shortest path and by the shortest time at once, why should it take one path over the other? It rather follows neither; it takes a path that has a more tangible advantage: the path it takes is the one with the least amount of action.[15]

Maupertuis essentially applies the rather vague and metaphysical claim that 'Nature, in the production of its effects, acts always by the simplest means'[16] to optics. Today one might object that a 'blind' nature, merely driven by efficient causes, does not 'know' which way the most efficient one is. But Maupertuis was open to teleological arguments in optics, like those proposed by Robert Boyle, who wanted to prove God's causal efficacy in the world. Boyle had renewed the idea of a teleological nature and was backed up by Newton. To propose a teleological interpretation of something usually described in corpuscular mechanics was therefore not completely unheard of at the time. The idea quoted above, that nature works in an *efficient* way, implies a teleological interpretation of natural events, because the efficiency of something can only be understood by taking into account the entire event from start to finish. Maupertuis saw fit to substitute *time* as a variable in Fermat's approach with a new variable, namely *action* as the entities 'saved' by nature, and he gives action the precise mathematical expression $\int vds$:

> Now I have to define what I mean by the quantity of action. When a body is transported from one point to another, a certain amount of action is required; this action depends on the speed of the body and on the distance it travels, but it is neither the speed nor the distance

15. 'En méditant profondément sur cette matière, j'ai pensé que la lumière, lorsqu'elle passe d'un milieu dans un autre, abandonnant déjà le chemin le plus court, qui est celui de la ligne droite, pouvoit bien aussi ne pas suivre celui du temps le plus prompt: en effet, quelle préférence devroit-il y avoir ici du temps sur l'espace? la lumière ne pouvant plus aller tout-à-la fois par le chemin le plus court, et par celui du temps le plus prompt, pour quoi iroit-elle plutôt par un de ces chemins que par l'autre? aussi ne suit-elle aucun des deux, elle prend une route qui a un avantage plus réel: le chemin qu'elle tient est celui par lequel la quantité d'action est la moindre'; 'Accord de différentes loix de la nature qui avoient jusqu'ici paru incompatibles' (1746), in *Œuvres de Maupertuis*, 4 vols (Lyon, Bruyset, 1768), vol.4, p.1-28 (16-17).
16. 'La Nature, dans la production de ses effets, agit toujours par les moyens les plus simples'; Maupertuis, 'Accord de différentes loix', p.12.

> taken separately. The quantity of action is as big as the speed of the body is big and as the path it has travelled is long, it is proportional to the sum of the distances travelled multiplied each by the speed at which they travelled. It is this quantity of action that is the real cost to nature here, which she manages to make as small as possible in the motion of light.[17]

However, his arguments for the necessity of the minimisation of action are obscure. They have the air of being more of a metaphysical postulate designed to fit empirical observations than being rigorous, scientific modelling. He simply postulates that there is some kind of variable that depends on velocity and distance. Some interpreters point out that, even though he found out a scientific 'truth', he came to it by pure guesswork.[18] This suggests that, instead of reconstructing a proper, well-formulated argument, we would have better luck if we were to trace the course that led Maupertuis to his discovery.

Maupertuis follows the Cartesian measure of movement as vds, but he refrains from using Leibniz's way of calculating forces as mv^2. He also makes use of Newton's idea that the pathway of light is the way that light-corpuscles travel, and thus generalises this optical formula by making it applicable to all physical bodies. Movement is determined by expending as little action as possible: 'Until some change happens in nature, the quantity of action necessary for such a change is the smallest possible.'[19] What is especially noteworthy here is how Maupertuis has made action the true cost to nature in general,

17. 'Il faut maintenant expliquer ce que j'entends par la quantité d'action. Lorsqu'un corps est porté d'un point à un autre, il faut pour cela une certaine action: cette action dépend de la vitesse qu'a le corps et de l'espace qu'il parcourt, mais elle n'est ni la vitesse ni l'espace pris séparément. La quantité d'action est d'autant plus grande que la vitesse du corps est plus grande, et que le chemin qu'il parcourt est plus long, elle est proportionnelle à la somme des espaces multipliez chacun par la vitesse avec laquelle le corps les parcourt. C'est cela, c'est cette quantité d'action qui est ici la vraie dépense de la Nature, et ce qu'elle ménage le plus qu'il est possible dans le mouvement de la lumière'; Maupertuis, 'Accord de différentes loix', p.17.
18. See Hermann von Helmholtz, 'Rede über die Entdeckungsgeschichte des Princips der kleinsten Action', in Adolf von Harnack, *Geschichte der Königlich Preussischen Akademie der Wissenschaften zu Berlin*, 3 vols (Berlin, 1900), vol.2, p.282-96 (294).
19. 'Lorsqu'il arrive quelque changement dans la Nature, la quantité d'action nécessaire pour ce changement est la plus petite qu'il soit possible'; Maupertuis, 'Recherche des lois du mouvement' (1746), in *Œuvres de Maupertuis*, vol.4, p.29-42 (36).

and has thus transferred it from optics to all natural events. He adheres to the traditional idea of light as consisting of corpuscular particles, and he also adopts Newton's arguments for inelastic collision (against Leibniz's idea of elastic collision). The metaphysical background to this reasoning is, as we now shall see, inspired by Leibniz's philosophy.[20] As Mary Terrall puts it: 'In pursuing this parallel [between physically extended bodies and light particles], Maupertuis's assumption of a Newtonian corpuscular theory of light ultimately led him to a decidedly non-Newtonian mechanics.'[21] Maupertuis highlights the organisational structure of the universe, which can be understood as governed by divine wisdom and causal efficiency alike:

> One cannot doubt that everything is governed by a supreme Being who has imposed forces on material objects that show his power, just as he has fated these objects to bring forth those effects that demonstrate his wisdom. The harmony between these two attributes is so perfect, that undoubtedly all the effects of Nature could be derived from each one taken separately. A blind and deterministic mechanics follows the plans of a perfectly clear and free Intellect. If our minds were sufficiently vast, we would also see the causes of all physical effects, either by studying the properties of material bodies or by studying what would be most suitable for them to do.[22]

Here he repeats the Leibnizian idea that we could, in principle, give both causal and final explanations for all events. This is important, as it allows him to apply a teleological explanation first and then assume that there must be an explanation by efficient causes. The economy of action derives from the divine wisdom by which creation itself is

20. For a more detailed account see Pulte, *Das Prinzip*, and Hartmut Hecht, 'Gemeinsame Denkmotive bei Leibniz und Maupertuis', in *Pierre Louis Moreau de Maupertuis: eine Bilanz nach 300 Jahren*, ed. Hartmut Hecht (Berlin, 1999), p.207-24.
21. Terrall, *The Man who flattened the earth*, p.177.
22. 'On ne peut pas douter que toutes choses ne soient réglées par un Etre suprême qui, pendant qu'il a imprimé à la matière des forces qui dénotent sa puissance, l'a destinée à exécuter des effets qui marquent sa sagesse; & l'harmonie de ces deux attributs est si parfaite, que sans doute tous les effets de la Nature se pourroient déduire de chacun pris séparément. Une méchanique aveugle & nécessaire suit les desseins de l'Intelligence la plus éclairée & la plus libre, & si notre esprit étoit assez vaste, il verroit également les causes des effets Physiques, soit en calculant les proprietés des corps, soit en recherchant ce qu'il y avoit de plus convenable à leur faire exécuter'; Maupertuis, 'Accord de différentes loix', p.21.

formed.[23] In this case, action is a means for establishing a certain natural order as an end: 'Nature, in the production of its effects, always acts by the most simple means.'[24] This is the reason why we never find any indication that Maupertuis might have considered it problematic to assume that a *teleological* principle derived from optics holds for all mechanics or dynamics: he adheres to the idea of a teleologically directed universe, following the plan of the creator – an idea which was prominently supported and comprehensively developed by Leibniz. Maupertuis makes it clear that this is an economical principle in which means are related in the most economical manner to their ends: 'But this "budget", this quantity of action that Nature minimises in the movement of light across different media, is it also minimised when it is reflected by opaque bodies and in its simple propagation? Yes, this quantity always has the smallest possible value.'[25]

This is, of course, a sign of the divine intelligence by which God has created the world. All events follow his wisdom. God has directed things according to his plan by forces, and Maupertuis merely iterates Leibniz's postulate of a harmony between final and efficient causes. There is a determined quantity of action in the world, and nature spends as little of this as possible. In the case of optics, nature 'saves' action in reflection and linear propagation so that light always takes the shortest and fastest path. In this way, Fermat's law is not a proper law by itself, but rather a consequence or expression of a more fundamental and universal law, namely the PLA. Maupertuis here takes it for granted that the laws of optics are basically just a special case of mechanics. For comparison, Newton also allows for the assumption of an economy of nature. Take, for example, his First Rule of Reasoning: '*No more causes of natural things should be admitted than are both true and sufficient to explain their phenomena.* As the philosophers say: Nature does nothing in vain, and more causes are in vain when fewer suffice. For nature

23. For the concept of an economy of the world and how it serves as a metaphysical ground for the derivation of the PLA, see Ansgar Lyssy, 'L'économie de la nature: Maupertuis et Euler sur le principe de moindre action', *Philosophiques* 42:1 (2015), p.31-51.
24. 'La Nature, dans la production de ses effets, agit toujours par les moyens les plus simples'; Maupertuis, 'Accord de différentes loix', p.12.
25. 'Mais ce fonds, cette quantité d'action que la Nature épargne dans le mouvement de la lumière à travers différens milieux, le ménage-t-elle également lorsqu'elle est réfléchie par des corps opaques et dans sa simple propagation? Oui, cette quantité est toûjours la plus petite qu'il est possible'; Maupertuis, 'Accord de différentes loix', p.19.

is simple and does not indulge in the luxury of superfluous causes.'[26] However, this is not an ontological principle, but rather a presupposition for scientific enquiry and formulating reasonable arguments. Newton makes the epistemic statement of ontological speculations clear in this famous passage, where he also rejects them from physics:

> I have not as yet been able to deduce from phenomena the reason for these properties of gravity, and I do not feign hypotheses. For whatever is not deduced from the phenomena must be called a hypothesis; and hypotheses, whether metaphysical or physical, or based on occult qualities, or mechanical, have no place in experimental philosophy. In this experimental philosophy, propositions are deduced from the phenomena and are made general by induction.[27]

For Maupertuis, who in this regard clearly sides with Leibniz, this ontological commitment plays quite a big role in the intersection of physics and metaphysics. In 1746, Maupertuis presented a paper called 'Derivation of the laws of motion and equilibrium from the attributes of God' at the Berlin Academy, which already pointed towards the foundation of mechanics in ideas of a rather theological nature. The published paper, as we have seen, bears the more modest title, 'The laws of motion and rest deduced from a metaphysical principle' (1746). Maupertuis applies the PLA to the direct impact of two bodies. He seems to be inspired by a certain understanding of Leibniz's concept of forces and the *vis viva* controversy. He adapts Leibniz's definition of force as mv^2 and, assuming that in any interaction of two bodies Σmv^2 is constant, he derives the change of velocity in both elastic bodies after collision as $m(v_0 - v_1)^2 + m'(v'_1 - v'_0)^2$ and it is this sum that is minimised in nature.[28] This is the action that is to be minimised, and it can be understood as an expression for the effect that forces have on other forces. It is clear that this discussion of laws takes place in a framework that is more inspired by Leibnizian dynamics than by Newton's rational mechanics, even though he does subscribe to the Newtonian concept of inelastic collisions.[29]

26. Isaac Newton, *The Principia: mathematical principles of natural philosophy*, translated by I. Bernard Cohen and Anne Whitman (Berkeley, CA, 1999), p.440 (original emphasis).
27. Newton, *The Principia*, p.589.
28. The notation is slightly abridged, since Maupertuis uses u_1 and u_2 for the velocities of both bodies after collision instead of v_1 and v'_1.
29. See Maupertuis, 'Les loix du mouvement et du repos déduites d'un principe

Maupertuis conceives, as was common in his time, the nature of light as the result of the movement of very small bodies. The laws of reflection and refraction are supposed to be part of the mechanics of rigid bodies, but it is somewhat unclear why the PLA has a different formula for the collision of bodies from the refraction of light. But, once again, the PLA is a metaphysical principle based on a specific understanding of nature as *creation*. It is one of the universal regularities by which we can discover the relationship between God and nature itself.[30] This means that, on the one hand, Maupertuis can deduce the PLA from his broad conception of an essentially economic creation and, on the other hand, simultaneously claim that the universal laws and regularities of the world prove the existence of God and the fact that the world is created by a supreme intelligence. He argues 'that there is a need to search for the proofs of God's existence in the general laws of nature; that the laws by which motion is conserved, distributed and destroyed are based on the attributes of a supreme intelligence.'[31] Maupertuis thus assumes on metaphysical grounds that the PLA holds for all physical events, but nonetheless his mathematics does not hold for dynamical systems in which one particle is subject to multiple forces. In short, the PLA serves as a regulative principle for scientific enquiry and metaphysical speculation alike: we understand physical effects and divine intentions alike in regard to the PLA. Let us now turn to Euler, who reworked Maupertuis's deduction of the PLA.

Euler

The mathematician and physicist Leonhard Euler was convinced that Maupertuis's account of the PLA was basically correct. Euler, even more than Maupertuis, emphasised that their 'indubitable' mechanics and empirical science should be the true source and foundation of 'imaginary' metaphysics: 'It would be absurd to think that mere imaginations could serve as foundations for the

---

métaphysique', in *Histoire de l'Académie royale des sciences et belles lettres* (1746; Berlin, Haude, 1748), p.267-94 (284).

30. See, for example, the discussion in Pulte, *Das Prinzip*, p.43.

31. 'Qu'il faut chercher les preuves de l'existence de Dieu dans les Loix génerales de la Nature. Que les Loix selon lequelles le Mouvement se conserve, se distribue et se détruit, sont fondées sur les attributs d'une suprême Intelligence'; Maupertuis, 'Les loix du mouvement', p.277.

real principles of mechanics.'[32] Mathematics and experience are the only true foundations of mechanics. Experience cannot serve as the basis for science, but neither can metaphysics, as it is rarely more than speculation and devoid of any relation to the *actual* world. His stance could be called 'neo-Euclidean', building on the assumption that axiomatic and self-evident truths can serve for a deduction of physical laws, in which mathematics would ideally be both ground and method.[33] Euler rejects Leibniz's monads as a mere fantasy; more importantly, he also rejects active forces, as these would conflict with the essential passivity of matter (see below). He also argues that mechanics can be axiomatically reduced to one single principle, namely Newton's second law, which defines f = ma.[34]

However, Euler combines this mathematical approach with some metaphysical presuppositions in his own deduction of the PLA. While Maupertuis was searching for universal principles that were obvious in optics, but equally valid in other scientific disciplines, Euler was rather looking for mathematical extrema.[35] In his 1744 treatise *A Method for finding curved lines enjoying properties of maximum or minimum*, Euler had developed a calculus of variations (i.e. a geometrical method to find curves – or other geometrical figures – in which a certain variable had a maximised or minimised value). Among such problems was the question: which curve would a moving body pass through, given a constant velocity, but subject to a constant force, which in this case was given by mass or gravitational force? Why does it not follow any other path? The calculus of variations offers an answer. To wit, the curve of a projectile can be determined by finding the property of maximum or minimum:

32. '[I]l seroit absurde de soutenir, que des pures imaginations pouvoient servir de fondement à des principes réels de la Mecanique'; Leonhard Euler, 'Réflexions sur l'espace et le tems', in *Histoire de l'Académie royale des sciences et belles lettres* (1748; Berlin, Haude and Spener, 1750), p.324-33 (326); as well as Christian Leduc's chapter in this volume. Translations of Euler's writings usually follow those accessible at http://eulerarchive.maa.org/ (last accessed 21 April 2022).
33. For a detailed analysis of Euler's method, see Pulte, *Das Prinzip*, and *Axiomatik und Empirie: eine wissenschaftstheoriegeschichtliche Untersuchung zur Mathematischen Naturphilosophie von Newton bis Neumann* (Darmstadt, 2005), ch.IV.3.
34. See Euler, 'Découverte d'un nouveau principe de mécanique', in *Histoire de l'Académie royale des sciences et belles lettres* (1750; Berlin, Haude and Spener, 1752), p.185-217. See also Pulte, *Axiomatik und Empirie*, p.170-86.
35. See J. Christiaan Boudri, *What was mechanical about mechanics: the concept of force between metaphysics and mechanics from Newton to Lagrange* (Dordrecht, 2002), p.156.

> Since all the effects of Nature obey some law of maximum or minimum, it cannot be denied that the curves described by projectiles under the influence of some forces will enjoy the same property of maximum or minimum. It seems less easy to define, *a priori*, using metaphysical principles, what this property is. But since, with the necessary application, it is possible to determine these curves by the direct method, it may be decided which is a maximum or a minimum.[36]

It is the curve s_1 that is determined by such a minimum:

> Let the mass of the projectile be M, and let its speed be v while being moved over an infinitesimal distance *ds*. The body will have a quantity of motion Mv that, when multiplied by the distance *ds*, will give M*vds*, the [quantity of] motion of the body integrated over the distance *ds*. Now I say that the curve [*linea*] thus drawn by the body is the curve (from among all other curves connecting the same endpoints) that minimises $\int$M*vds* or, provided that M is constant, $\int$v*ds*.[37]

Euler realised that Maupertuis's momentary action is valid only for infinitesimal moments and that consequently the action along an arbitrary infinitesimal section of *s* has a minimal value. One

36. 'Quoniam omnes naturae effectus sequuntur quandam maximi minimive legem; dubium est nullum, quin in lineis curvis, quas corpora projecta, si a viribus quibuscunque sollicitentur, describunt, quaepiam maximi minimive proprietas locum habeat. Quaenam autem sit ista proprietas, ex principiis metaphysicis a priori definire non tam facile videtur; cum autem has ipsa curvas, ope Methodi directae, determinare liceat; hinc, debita adhibita attentione, id ipsum, quod in istis curvis est maximum vel minimum, concludi poterit'; Leonhard Euler, *Methodus inveniendi lineas curvas maximi minimive proprietate gaudentes, additamentum* 2: *De motu projectorum in medio non resistente* (Lausanne and Geneva, Bousquet, 1744), p.311-20 (311). The translation follows René Dugas, *A History of mechanics* (Mineola, NY, 2012), p.273. In this *additamentum*, Euler strives to derive the trajectory of a body moving within a non-resisting medium, but he incorporates an external force that constantly acts on the body, effectively deriving some of the laws that define one single planetary orbit (without the gravitational interference of other bodies). For a more detailed reconstruction, see also Rojo and Bloch, *The Principle of least action*, p.65-67.
37. 'Sit massa corporis projecti = M, ejusque, dum spatiolum = ds emetitur, celeritas debita altitudini *v*; erit quantitas motus corporis in hoc loco = Mv; quae per ipsum spatiolum ds multiplicata, dabit Mdsv motum corporis collectivum per spatiolum ds. Jam dico lineam a corpore descriptam ita fore comparatam, ut, inter omnes alias lineas iisdem terminis contentas, sit $\int$Mdsv, seu, ob M constans, $\int$dsv minimum.' Euler, *Methodus inveniendi, additamentum* 2, p.311-12.

needs to integrate this momentary action over the entire path of the particle. Action, thus, is not a property of things or events, but rather a property of the geometrical model that describes the unfolding of an event over time. Euler's account differs from Maupertuis's in the way it is deduced in a proper mathematical manner.

The resulting curve is determined by the PLA as the sum of all infinitesimal sections and has a minimal value in comparison to all other possible curves. Action is formalised as $I = \int mv\,ds$ and can also be written as $I = \int mv^2 dt$, as $v = ds/dt$, and therefore $v\,dt = ds$. Here we encounter again the Leibnizian force (*vis viva*) $f = mv^2$, which is conceived as the causal effect of forces in regard to velocity in a certain point. With this in mind, we can also say that the resulting curve is determined by the PLA as the sum of all *momentary* expressions of force over time, and is minimised in comparison to all other curves. Of course, Euler presupposes not only that curves and all types of change are continuous, but also that all possible curves form a spectrum of continuity.

He further recognised that this was the same PLA as formulated by Maupertuis. But it expressed the movement of a body under forces expressed by geometrical means, while Maupertuis was restricted to massless bodies without forces. Consequently, Euler had to add another condition to make his formula valid, namely that the forces involved had to be constant – expressing it in a modern way: that the energy involved remained constant. This new definition of *action* makes the PLA directly relevant to our understanding of causal processes, and allows for an interpretation in terms of causal relations: 'His [Maupertuis's] principle is that of the least possible action; by which he means, that in all the changes which happen in nature, the cause which produces them, is the least that can be.'[38]

The actual trajectory of the particle, in contrast to all possible trajectories, can be found out both with a direct and with an indirect method. The so-called 'direct' approach consists in relating the different moments of a movement to the forces involved, explaining

38. 'Son principe est celui de la moindre action, par lequel il prétend que dans tous les changemens, qui arrivent dans la nature, l'action qui les opère est toujours la plus petite qui soit possible'; Leonhard Euler, *Lettres à une princesse d'Allemagne sur divers sujets de physique et de philosophie*, 2nd edn, 2 vols (Leipzig, Steidel, 1770), vol.1, letter LXXVIII, p.331. Translations for this text follow *Letters of Euler on different subjects in physics and philosophy addressed to a German princess*, translated by Henry Hunter, 2nd edn, 2 vols (London, 1802), vol.1, p.302-303.

every consequent change by Newtonian forces. The indirect approach is geometrical: it is based on the calculus of variations, and it defines the movement over the course of time in regard to a variable, here *action*, which is minimised over the entire course. It is considered a teleological principle, because here the entire course of events is considered, but it is also thought to be a determining principle from the very outset. Both methods can be equally employed, depending on the variables given in regard to a specific situation:

> From this one can see that there has to be a double method for resolving the problems of mechanics; one is the direct method, which is grounded in the laws of equilibrium, or movement; but the other is that which I am going to speak about, where the solution can be found by the means of the methods *de maximis et minimis*, if you know the formula, which has to be a *maximum* or a *minimum*. The first one provides the solution in determining the effect of efficient causes; and the other one has final causes in view.[39]

Here we see that Euler refers to Leibniz's doctrine that every causal relation can be deduced from efficient and final causes alike. Despite his attack on metaphysics, especially on Leibnizian metaphysics, Euler does not distinguish in clear terms between metaphysics and physics, which makes his arguments shift seamlessly between the one and the other. He criticises Leibniz for deriving his physical forces from primitive active forces, which are embodied in infinitely small parts of matter;[40] yet he clearly accepts that the notion of force needs a

39. 'Par là on voit qu'il doit y avoir une double méthode de resoudre les problèmes de Mécanique; l'une est la méthode directe, qui est fondée sur les loix de l'équilibre, ou du mouvement; mais l'autre est celle dont je viens de parler, où sachant la formule, qui doit être un *maximum*, ou un *minimum*, la solution se fait par le moyen de la méthode *de maximis et minimis*. La première fournit la solution en déterminant l'effet par les causes efficientes; or l'autre a en vue les causes finales'; Euler, 'Recherches sur les plus grands et plus petits qui se trouvent dans les actions des forces', in *Histoire de l'Académie royale des sciences et belles lettres* (1750; Berlin, Haude and Spener, 1752), p.149-88 (151).
40. See Euler, *Gedancken von den Elementen der Cörper* (Berlin, Haude and Spener, 1746); reprinted in *Opera omnia*, series 3, vol.2 (Basel, 1942), p.347-66. The entire argument is a straw-man argument, since Euler ascribes to Leibniz the theory that each body consists of a finite and definite number of infinitely small particles, which are monads (see part II, §8). This, however, misrepresents Leibniz's sophisticated mereology of physical bodies and unextended substances, as the number of monads that constitute the physical reality of any body is actually infinite and they are not spatially contained within it. For a

metaphysical background to be intelligible. He grounds it then in the 'essence of bodies', in which the 'reason' of the force is embodied[41] – this idea that force can have an internal and external reason, which he proposes elsewhere,[42] also has a distinctly Leibnizian tone to it – for Leibniz, force follows the individual principle of the substance, nominally defined by the complete notion of the subject, while also, by means of the pre-established harmony, adhering to contextual and external factors *as if* it were mechanically directed by them as well. It is important to him that this essence of bodies is nothing but matter itself, and that all apparently active forces can be reduced to passive forces, which he first identifies as inertia,[43] and later as impenetrability.[44] In this way he apparently strives to eliminate irreducibly active forces from his physics, which he often associates with Leibniz's doctrine of monads.

Euler decides to ignore speculations about the nature of forces in favour of enquiring into their effects. The idea guiding his research is that bodies obey forces as little as possible.[45] Forces are real but *derivative* entities. They result from underlying substances. This also accounts for the fact, again in an obviously Leibnizian manner, that there are final causes embedded in natural forces. In one of his later writings, Euler makes it clear that final causes are real and embedded in the metaphysical or substantial nature of bodies: the success of the enquiry into movements by final causes is 'founded in the nature of bodies themselves'.[46] There is quite a lot

comprehensive account of this complicated mereology and the discussion about how it should best be understood, see, for example, Stefano di Bella, *The Science of the individual: Leibniz's ontology of individual substance* (Dordrecht, 2005); Glenn Hartz, *Leibniz's final system: monads, matter and animals* (London, 2007); Ansgar Lyssy, *Kausalität und Teleologie bei G. W. Leibniz* (Stuttgart, 2016), parts III-V; and Ohad Nachtomy, 'Leibniz on nested individuals', *British journal for the history of philosophy* 15:4 (2007), p.709-28.

41. 'Da nun ein jeglicher Cörper eine solche Kraft besitzet in seinem Zustande zu verharren, so muß der Grund derselben in dem Wesen der Cörper enthalten sein'; Euler, *Gedancken von den Elementen*, part II, §15.
42. For example, in Euler, *Theoria motus corporum solidorum seu rigidorum* (1765), in *Opera omnia*, series 2, vol.3 (Basel, 1948), ch.II.
43. In Euler, *Gedancken von den Elementen*, part II, §10.
44. In Euler, 'Recherches sur l'origine des forces', in *Histoire de l'Académie royale des sciences et belles lettres* (1750; Berlin, Haude and Spener, 1752), p.419-47.
45. Euler, *Methodus inveniendi, additamentum* 2, p.320; see Dugas, *A History of mechanics*, p.274.
46. '[f]ondé dans la nature même des corps'; Euler, *Lettres à une princesse*, vol.1, letter LXXVIII, p.331.

of textual evidence that Euler's justification of the PLA draws from two sources or 'background' ideas: first, from the mathematical attempts to determine the unequivocal minimum of a curve, say, a chain fixated on two points or the trajectory of a moving body; and, second, from a metaphysical background easily identifiable as Leibniz's philosophy. This consists in the same belief in the two reigns of final and efficient causes as has been shown in regard to Maupertuis, which in Euler's writings connect the economy of the world to theological ideas. This is evident in one of the *Letters to a German princess*:

> If we consider the order and economy of the world in this point of view [in regard to the non-arbitrary decree of God], nothing can be more perfectly disposed for the attainment of this great end [human felicity]. All events, the calamities themselves which we undergo, are the most suitable means for conducting us to true happiness; and in this respect it may be with truth affirmed, that this is indeed the best world possible, as everything in it concurs to promote our eternal salvation. [...] [N]othing befalls me by chance; but [...] all events are directed by Providence.[47]

Here it becomes apparent that Euler, even though an adherent of mechanism and the explanation of all events by mathematical science, nonetheless advances the thesis that there is a moral, or normative aspect to be found in all events. Another piece of evidence points us directly to the economy of nature as proposed by Aristotle and Leibniz:

> The earliest philosophers had already recognised that nature does nothing in vain and this relates perfectly to [the question of] least expense. For if nature were to employ superfluous expense, it would clearly involve nature doing something in vain. Aristotle already makes reference to this doctrine, and it seems that he has taken this idea from those who preceded him rather than envisioning it himself. The proposition has subsequently had such enormous influence in the

47. 'Si nous regardons sur ce pied l'arrangement et l'administration de ce monde, tout ne saurait être mieux disposé pour ce grand but [la félicité des hommes]. Tous les événements et même les adversités que nous éprouvons, sont les moyens les plus propres pour nous conduire à notre vrai bonheur: et à cet égard on peut dire que ce monde est effectivement le meilleur, puisque tout y concourt à opérer notre salut. [...] [I]l ne m'arrive rien dans ce monde par hasard, et [...] tous les événements en sont dirigés par une providence'; Euler, *Lettres à une princesse*, vol.2, letter CXIII, p.168.

> Schools that it was regarded as one of the first precepts of philosophy until Descartes ultimately dared to reject it.[48]

Euler here assumes that this conception of an economy of nature has been a main principle of philosophy from Aristotle to Descartes, and he relates this philosophical idea directly to his idea of least expense. Even though he occasionally displays a strong aversion to metaphysical speculation, especially against the ideas of Leibniz and Wolff, metaphysical speculation about the overall economy of the world does nonetheless help Euler, just as Maupertuis, as an explanatory guide for his scientific research. The economy of nature is articulated in the more concrete striving of nature for a *minimum*, or, depending on which unities or aspects the scientist focuses on, a maximum. As Joseph Petzoldt has claimed,[49] the PLA provides Euler with a mathematical mode of access for understanding nature as *determined* in comparison to all other possibilities. This is only possible by the mathematical approach that looks for mathematical extremes, not for metaphysical principles, as Maupertuis had done. We even find such a statement in Euler's own writing:

> For since the plan of the universe is the most perfect possible and the work of the wisest possible creator, nothing happens which has not some maximal or minimal property, and therefore there is no possible doubt that all effects in the world can be *determined* equally well from final causes using the method of maxima and minima, or from the efficient causes themselves.[50]

48. '*Car les plus anciens Philosophes avoient* déjà reconnu, que la nature ne faisoit rien en vain, ce qui s'accorde parfaitement avec la moindre dépense; car si la Nature employoit des dépenses superfluës, il n'y a pas de doute qu'elle ne fit quelque chose en vain. Aristote fait déjà souvent mention de ce dogme, et paroit l'avoir plutôt pris de ceux qui l'avoient précedé, que l'avoir imaginé lui-même. La proposition a fait ensuite un si grand progrés dans les Ecoles, qu'on l'a regardée comme un des premiers préceptes de la Philosophie, jusqu'à ce qu'enfin Descartes a osé la rejetter'; Euler, 'Sur le principe de la moindre action', in *Histoire de l'Académie royale des sciences et belles lettres* (1751; Berlin, Haude and Spener, 1753), p.199-218 (199). Translation follows http://eulerarchive.maa.org/pages/E198.html (last accessed 21 April 2022).
49. See Joseph Petzoldt, 'Maxima, Minima und Oekonomie', *Vierteljahresschrift für wissenschaftliche Philosophie* 14 (1890), p.206-39, 354-66 and 417-42, esp. ch.I.
50. 'Cum enim Mundi universi fabrica sit perfectissima atque a Creatore sapientissimo absoluta, nihil omnino in mundo contingit, in quo non maximi minimive ratio quaepiam eluceat; quamobrem dubium prorsus est nullum, quin omnes Mundi effectus ex causis finalibus ope Methodi maximorum et

The idea expressed is the following: any possible curve that would *not* satisfy the PLA would have a 'twin' (i.e. a symmetrical and corresponding trajectory with the same values). The PLA helps to determine the one trajectory that is unequivocally determined (i.e. that has no such 'twin'). This is precisely the idea that Leibniz developed in the *Tentamen anagogicum*, written *c.*1696, however not published before the nineteenth century. The similarity between Leibniz's and Euler's approaches may be due to both searching for a mathematical equivalent of the Principle of sufficient reason.[51] However, Leibniz developed more general versions of the argument elsewhere as well, for example in the 'Unicum opticae, catoptricae et dioptricae principium' ('A unitary principle of optics, catoptrics and dioptrics'), which was published in the *Acta eruditorum* in 1682, where he argues that light travels along the path that is unique against all other possible paths in regard to 'ease', and that this uniqueness is an expression of God to create the most perfect world possible.[52] Leibniz also picks up on this in his famous *Discourse on metaphysics* (1686), where he writes that it is 'God's decree always to produce his effect by the easiest and most determined ways'.[53] Euler would most likely have known these ideas.

Here we also find displayed another metaphysical, especially Leibnizian, influence: each event can be equally understood and explained by mechanistic as by final causes. This, of course, is due to the fact that the world is a creation by God who has outlined a divine plan for everything that has happened or will ever happen. For Leibniz, though, the determining principle is the Principle of

---

minimorum aeque feliciter *determinari* queant, atque ex ipsis causis efficientibus'; Euler, *Methodus inveniendi, additamentum* 1: *De curvis elasticis* (Lausanne and Geneva, Bousquet, 1744), p.245-310 (245, my emphasis). Translation follows Philip E. B. Jourdain, 'Maupertuis and the Principle of least action', *The Monist* 22 (1912), p.414-59 (435).

51. However, this only holds for idealised curves. As soon as friction is taken into account, the *actual* pathway of a projectile within a gravitational field would be an *asymmetrical* parabola. For such an asymmetrical parabola, there is a symmetrical 'twin' for which action ($I = \int mv\,ds$) has the same value.

52. G. W. Leibniz, 'Unicum opticae, catoptricae et dioptricae principium', *Acta eruditorum* (June 1682), p.185-90.

53. Leibniz, *Discourse on metaphysics*, in *Philosophische Schriften*, ed. C. I. Gerhardt, vol.4 (Hildesheim, 1978), p.427-63 (447); reprinted in *Sämtliche Schriften und Briefe*, ed. Berlin-Brandenburgische Akademie der Wissenschaften, vol.VI.4 (Berlin, 1999), p.1529-88 (1563). See *G. W. Leibniz: philosophical essays*, ed. and translated by Roger Ariew and Daniel Garber (Indianapolis, IN, 1989), p.54.

sufficient reason, which postulates that nothing happens without a determining reason, a reason that can also be understood as a cause.

Yet the mathematical approach was more favourable to Euler, who still strives to reject metaphysical speculation from his physics, even though he cannot get rid of it entirely. He used this advantage of a purely mathematical approach to criticise Maupertuis's attempt to relate the PLA to the actions of forces: 'the determination of the quantity of action bears no relation to any force whatsoever.'[54] We can formulate action as $I = \int mv\,ds$ and mass, velocity and path are here the only relevant factors – but his *metaphysical* idea of the equivalence of explanations by efficient and final causes leads him to assume that this would correspond to Newton's forces. The advantage of the PLA is that Maupertuis's unresolved question of how much force is required to give a body a certain velocity can be simply ignored, at least within the mathematical approach. The determination of movement by forces (e.g. *vis viva*) requires a metaphysical background Euler does not want to deal with.

The same holds for all other non-conservative forces. As such, Euler's *mathematical* version of the PLA only holds for moving projectiles subject to no or only one constant force, and then only under idealised conditions, as it would allow for two possible solutions. The *direct* and *non-teleological* approach would unequivocally determine this as the only possible solution. Of course, it should be noted, Euler is well aware of the problem that his deduction of the PLA holds only for conservative forces, and he assumes that a further elaboration of *metaphysics* will lead to a greater clarity of his solution.[55] Consequently, he rather proposes to rework the distinction between conservative and non-conservative forces instead of restraining the general validity of the PLA, which he assumes on purely metaphysical grounds.

The major difference between Maupertuis and Euler is that the latter extends the PLA to dynamics, and supplies it with a theory of forces with which he intends to show how all types of forces can be reduced to one (see below), which is clearly meant to justify the generalisation of the PLA. This is not a minor detail. This claim is supposed to support the validity of the PLA for *all* kinds of physical events: since there is only one type of force, all kinds of action can be

54. 'La determination de la quantité d'action n'a egard à aucune force'; Euler to Maupertuis (24 May 1746), in *Opera omnia*, series 4.a, vol.6 (Basel, 1986), p.64.
55. See Euler, *Methodus inveniendi, additamentum* 2, p.320.

modelled after the trajectory of one body, the collision of two bodies and the parallelogram of forces.

Leibniz had postulated that there are two fundamental properties of matter, namely inertia and impenetrability. Both correspond to two different types of forces, that is, derivative and primitive forces. Leibniz makes it sufficiently clear that only one of these possesses substantial reality and the other one, the derivative force, is only a phenomenon. However, in a text entitled 'Recherches sur l'origine des forces' ('Research on the origin of forces', 1752), Euler argues against Leibniz by maintaining that there is only one fundamental property of matter, impenetrability. Euler then goes on to derive extension, mobility and inertia from impenetrability, and thus to ground the different physical aspects of matter in one single property. This property is then to be conceived as an irreducible and thus *substantial* property of matter itself, and can consequently be understood as a *metaphysical* concept.[56]

To summarise the argument very briefly, it goes as follows: without extension, impenetrability is meaningless, for a thing that has no extension cannot occupy a place. Thus, any impenetrable object is necessarily extended.[57] Due to its impenetrability, an object occupies a place and thus can potentially change place – in other words, impenetrability implies mobility. If an object is mobile, then inertia also has to be attributed to it, as inertia prevents changes from occurring without an external, sufficient reason: 'In the impact between bodies, their impenetrability produces only the minimum force required to protect them from penetration [...] and no doubt it is this fact that is the basis of the general principle that all changes in the world are produced at the least expense possible, or with the smallest forces capable of producing the effect concerned.'[58] Both force and action

56. See Pulte, *Das Prinzip*, p.23-24 and 110-11.

57. See the posthumously published *Anleitung zur Naturlehre*, in *Opera omnia*, series 3, vol.1 (Leipzig, 1926), p.16-180 (§35-39, p.75); and *Theoria motus corporum solidorum seu rigidorum*, §131-35. See also Dieter Suisky, 'Euler's mechanics as a foundation of quantum mechanics', in *Leonhard Euler: life, work and legacy*, ed. Robert E. Bradley and Charles Edward Sandifer (Amsterdam, 2007), p.503-25; Stephen Gaukroger, 'The metaphysics of impenetrability: Euler's conception of force', *British journal for the history of science* 15 (1982), p.132-54.

58. 'Ainsi dans le choc des corps leur impénétrabilité ne fournit toujours que la plus petite force, qui est capable de les garantir de la pénétration; et c'est sans doute sur cette circonstance, qu'est fondé ce principe si général, que tous les changemens au monde sont produits aux moindres dépens qu'il est possible, ou

are subject to the same minimum principle. All forces and relevant physical properties are effects of impenetrability. This should be best understood as an attempt to eliminate the unwelcome metaphysics of force from his physics. Euler outlined an order or hierarchy of the characteristics of physical bodies and did so expressly in his later *Anleitung zur Naturlehre* (published posthumously in 1862, but most likely written between 1744 and 1748): 'Impenetrability does by itself already include extension and mobility and consequently also inertia in itself. Thus, if one ascribes impenetrability to a body, one also has to ascribe it the other attributes too.'[59]

However, this has important consequences for the interpretation of the PLA within the context of Euler's physics: all force can only be attributed to bodies in so far as they collide or could potentially collide. Euler accepts no action-at-a-distance forces, so all action results from contact. Thus, the theory of contact forces is supposed to unite all physical phenomena. In the end, all motion results from uniform motion, and all kinds of bodies are composed of bodies of infinitesimal magnitude.[60] This allows him to use collision as a paradigm for any type of change,[61] and therefore to understand forces by means of the well-defined mechanical account of collision. J. Christiaan Boudri highlights the importance of this reduction:

> Euler's foundation of the force of impact shows even more strongly than Maupertuis's argument that the final cause and efficient cause have the same metaphysical basis. Where for Maupertuis this basis was the oneness of God, for Euler the reference to God is as it

---

avec les plus petites forces, qui sont capables de cet effet'; Euler, 'Recherches sur l'origine des forces', p.431-32. Translation follows Boudri, *What was mechanical*, p.168.

59. 'Die Undurchdringlichkeit schliesset für sich schon die Ausdehnung und Beweglichkeit und folglich auch die Standhaftigkeit in sich. Wenn man also dem Körper die Undurchdringlichkeit zueignet, so muss man ihm auch die übrigen Eigenschaften zuschreiben'; Leonhard Euler, *Anleitung zur Naturlehre*, in *Opera postuma mathematica et physica*, ed. Paulus Henricus Fuss and Nicolaus Fuss, 2 vols (St Petersburg, 1862), vol.2, p.449-560 (472). See also 'So nehmen alle Kräfte, welche die Veränderungen in der Welt hervorbringen aus der Undurchdringlichkeit der Körper ihren Ursprung, und daher sind in der körperlichen Welt keine anderen Kräfte anzutreffen, als die, welche aus der Undruchdringlichkeit der Körper entstehen' (Euler, *Anleitung*, p.474).
60. Euler, *Mechanica sive motus scientia analytice exposita* (St Petersburg, Petropoli, 1736), part I, *propositio* III; see Dieter Suisky, *Euler as physicist* (Dordrecht, 2009), p.134.
61. See Euler, *Lettres à une princesse*, vol.1, letter LXXIX.

> were reflected onto matter itself. [...] What is concealed in his foundation [...] is that the minimality of the integral of action does not say anything about the separate material substances, but about their structure. In the end, the forces of impact are not derived from the characteristics of the material substances, but from a principle of material structures: that the interactive effects are as small and as direct as possible.[62]

Despite this metaphysical dimension of Euler's concept of force, he understands the PLA as justified not by an *a priori* deduction, but with an *a posteriori* verification, since the conservation of energy or of forces had to be presupposed. Yet it was considered a basic principle, as it was to be found and confirmed in experience. For Maupertuis and Euler, *action*, nonetheless, is the measure of God's own efficiency and not justifiable by scientific enquiry alone. It needs a confirmation that effectively includes experience and metaphysical speculation alike. For both, material substance cannot be a starting point of philosophy anymore. The concept of force is based on experience, and the idea of a universal regularity is based on divine wisdom, which together allow them to assume the equivalence of final and efficient causes within a metaphysical framework. For both, the PLA is a regularity principle to which other laws and theorems have to conform. It has become a universal principle independent from any specific understanding of forces. It is not force anymore that defines what is physically possible or not, but to a great extent the PLA itself. In this sense, the PLA does not substitute any broader conception of a generally *efficient* nature, but gives it a precise and applicable physical interpretation. As a result, this metaphysical conception of nature has been transformed from a metaphysical concept of force to a way of conceiving of force by means of its *quantifiable effect*. The idea that force has to be grounded in a material *substance* was transformed into a project of grounding force in the *structure* of matter. Geometrical deduction has taken the place of former attempts at giving force a metaphysical foundation.

The Principle of least action in context

Speaking very broadly, it is clear that Leibnizian ideas – or at least those ideas that found their most recent and elaborate articulation in Leibniz, such as the economy of nature, the equivalence of final and

62. Boudri, *What was mechanical*, p.169-70.

efficient causes, the conception of force as mv^2 – play an important role in the deductions given by Maupertuis and Euler. They serve as a 'grounding' for dynamics in Maupertuis and Euler, even though the latter is reluctant to admit it and instead argues that mechanics could be grounded in merely axiomatic principles that have no metaphysical dimension. So, while Euler designs a Euclidean outlook of deductive mathematical science, his arguments are still infused with some metaphysical ideas. The problem with metaphysics, for Euler, is its vagueness and uncertainty, which can only be compensated for by mathematical deduction and formalisation. In a sense, metaphysical principles may serve as axioms, even though they lack the clarity of mathematical axioms, unless they are explicitly formalised in mathematical terms. These metaphysical principles have a double function. First, they help us clarify the 'why' of the more specific laws of nature, quite in contrast to Newtonianism, which explicitly confines itself to the description of 'how' things are correlated. Second, they serve as guidelines both for our experimental scientific enquiries, in which we are enticed to look for minimised values, and for the theoretical unification of science, as they help us relate different laws to each other. Maupertuis had illustrated the latter point in his paper 'Accord de différentes loix', in which he had shown that the PLA could be used to unify the different formulas for elastic and inelastic collisions. In a sense, metaphysics had lost its 'superior' position as the starting point of all science, as Descartes had envisioned; instead it had now become the 'most general' of all sciences. While Leibniz had envisioned that we could (in theory, though not always in practice) deduce all natural laws from a couple of metaphysical principles, Euler – and to a certain degree also Maupertuis – rejects this overtly strict approach by expressing scepticism towards all metaphysical ideas that could not be reasonably formalised. Only by means of mathematical formulas could metaphysical ideas become relevant to science, at least in principle, even though their actual methodology shows a different approach.

But as soon as we look at the details, differences emerge. Maupertuis did accept Leibniz's notion of force and did not make much effort to get rid of the metaphysical ballast attached to it. Euler, by contrast, wanted to avoid a Leibnizian dynamics in favour of rational mechanics, at best modelled on Euclid's view of an axiomatic science.[63] He instead proposes a reduction of force to matter, which is essentially

63. See Pulte, *Axiomatik und Empirie*, ch.4.

a Cartesian position, and in which the *res extensa* is considered passive, while Euler still acknowledges an active but immaterial *res cogitans*.[64] His conception(s) of passive force as grounded in the 'essence of matter', first as inertia, then as impenetrability, however, lack(s) any suggestion as to how we should understand the metaphysically loaded conception of 'essence' beyond simply conceiving of it as inertia or impenetrability. Maupertuis was much more open to formulate hypotheses based on experience, while Euler tried to keep both hypothesis and experience out of science and looked for a merely deductive method. Finally, Maupertuis acknowledges room for an ontology of nature under the name 'cosmology' – something that for Euler would again be too speculative and thus too imprecise to count as science, as Euler's emphasis on mathematical formalisation and his Euclidean method leaves little room for ontological speculations. It is also clear that this mathematics-based approach cannot be used for all topics, and neither Maupertuis nor Euler attempts this. When it comes to questions concerning the existence of God[65] or the relationship between the mind and the body,[66] both Maupertuis and Euler turn to rather traditional metaphysical approaches. So the description above only holds for what would be called natural philosophy or cosmology (i.e. those parts of metaphysics that deal with the nature of bodies and temporal change). But, with these reservations, the method outlined above fits both with the brief suggestion of the editor of the *Histoire* cited near the beginning of this chapter and with Maupertuis's descriptions of the duties of the academicians.[67]

How does this hold up within the context of the Academy of Berlin and its journal, the *Histoire*, in general?[68] This is difficult to answer, as there are few other papers one could compare to the approach developed by Maupertuis and Euler (i.e. few other papers that explicitly try to relate general mechanical debates to metaphysics). In 1765, six years after Maupertuis's death, Jean Bernard Merian published a work called 'Discours sur la métaphysique' ('Discourse on metaphysics'), in which he sketches a 'division of labour' between metaphysics and

64. See Pulte, *Das Prinzip*, part B. For Euler's conception of the active soul, see Euler, *Lettres à une princesse*, vol.2, letters LXXX-LXXXII.
65. See, for example, Maupertuis, 'Examen philosophique', p.389-424.
66. See, for example, Euler, *Lettres à une princesse*, vol.2, letter LXXXI.
67. See the appendix of this volume.
68. For a more general account of the attitudes towards science and metaphysics in the Berlin Academy, see Mary Terrall, 'The culture of science in Frederick the Great's Berlin', *History of science* 28 (1990), p.333-64.

physics, with physics discovering the facts and metaphysics supporting these facts with proofs and analysing them. Thus, science and metaphysics are interrelated, even though Merian does not emphasise mathematics as much as Maupertuis and Euler.[69] As Christian Leduc suggests in a recent paper,[70] it is Nicolas Béguelin who might be closest to both, though he was a more genuine follower of Leibniz and Wolff. In his 1757 paper 'Les premiers principes de la métaphysique' ('The first principles of metaphysics'), Béguelin argues that science has to be led by metaphysical principles, but he also argues that metaphysics can never have any mathematical certainty.[71] Given the absence of other publications within the field of natural philosophy, which at least implicitly deal with the relation between mathematics, physics and metaphysics, it is hard to compare Maupertuis and Euler to other members.

However, the significance of their approach can best be illustrated with a different example. The PLA was criticised by a French mathematician called Patrick d'Arcy (1725-1779), who wrote a text called 'Réflexions sur le principe de moindre action de M. de Maupertuis' ('Reflections on the Principle of least action of M. Maupertuis'), published first in 1749 in the *Mémoires de l'Académie des sciences de Paris*.[72] Without getting into the details of the nature of this criticism, it should be noted that it was much more Newtonian than the version of Maupertuis and Euler, and thus fittingly published by the Academy in Paris, not the Berlin Academy. While the texts were usually first presented at the Paris Academy and then submitted for publication,

69. To be fair, Merian does not give a precise definition of his conception of *physique*, and it seems to rather mean a broadly construed natural philosophy, as it includes such diverse topics as mechanism, electricity, magnetism, light and also the generation and organisation of living bodies. However, he also includes therein a narrower conception of physics as an investigation into the universal laws of nature by means of geometry, which would make his *physique* at least partially comparable to Maupertuis's and Euler's approach. See Merian, 'Discours sur la métaphysique', in *Histoire de l'Académie royale des sciences et belles lettres* (1765; Berlin, Haude and Spener, 1767), p.450-74. See also Christian Leduc, 'La métaphysique de la nature à l'Académie de Berlin', *Philosophiques* 42:1 (2015), p.11-30 (29).
70. Leduc, 'La métaphysique de la nature', p.30.
71. Nicolas Béguelin, 'Les premiers principes de la métaphysique', in *Histoire de l'Académie royale des sciences et belles lettres* (1755; Berlin, Haude and Spener, 1757), p.405-23.
72. Patrick d'Arcy, 'Réflexions sur le principe de moindre action de M. de Maupertuis', *Mémoires de l'Académie des sciences de Paris* (1749/1753), p.531-38.

the text written by d'Arcy was published in advance and presented only three years later. This urgency indicates that the publication of this particular text was a matter of high priority for the Paris Academy, which apparently was somewhat startled by the impact the PLA had on the scientific community. Maupertuis answered this text with a 'Réponse à un mémoire de M. d'Arcy' ('Response to a memoir by M. d'Arcy'), published again in the *Histoire* of 1752.[73] This, in turn, motivated d'Arcy to write another text called 'Réplique à un mémoire de M. de Maupertuis' ('Rejoinder to a memoir by M. Maupertuis'), again published by the Paris Academy in 1752.[74] As Maupertuis was too ill or too disinterested to answer, and since Euler had apparently already moved on to other topics, another mathematician named Louis Bertrand (1731-1812) chimed in to the debate with a piece called 'Examen des réflexions de M. le chevalier d'Arcy sur le principe de la moindre action' ('Examination of the reflections of M. d'Arcy on the Principle of least action'), which was first read on 3 October 1754, then published in winter 1757 in the *Histoire* of the Berlin Academy. This paper failed to create any resonance, an indication that the debate between the two academies about the PLA was already over. While Maupertuis and Euler remain without notable peers in the Berlin Academy, at least until the arrival of D'Alembert, one may reasonably assume that the actual frontlines were drawn between the different projects of the Berlin Academy and the academy in Paris. But this would have to be the topic of another chapter.

73. Maupertuis, 'Réponse à un mémoire de M. d'Arcy inséré dans le volume de l'Académie royale de Paris pour l'année 1749', in *Histoire de l'Académie royale des sciences et belles lettres* (1752; Berlin, Haude and Spener, 1754), p.293-98. It is worth mentioning that both Maupertuis's essay 'Les loix du mouvement' and his 'Réponse à un mémoire de M. D'Arcy' were published in the class of speculative philosophy, instead of being published under experimental philosophy. This reinforces the idea that this was a genuine metaphysical discussion for Maupertuis, rather than one which could be decided by experience, thereby contributing to working out a metaphysical underpinning for experimental science.

74. Patrick d'Arcy, 'Réplique à un mémoire de M. de Maupertuis', *Mémoires de l'Académie des sciences de Paris* (1752/1756), p.503-19.

Euler and the reflexive origin of the idea of space

CHRISTIAN LEDUC

Leonhard Euler engages in one of the most important scientific and metaphysical controversies in the eighteenth century: the debate on the nature of space and time. As is well known, the discussion focuses on two main positions: the Leibnizian view, according to which space and time are relational and thus phenomenal determinations of things; and the Newtonian view, which claims that space and time are instead absolute entities of the world. Most of the participants in the ongoing debate, including Clarke, Raphson, Berkeley, Wolff, Du Châtelet and Kant, embrace either the former or the latter position, each adapting it to their own theoretical framework. For his part, Euler supports Newton's thesis, especially concerning the characteristics of space. Euler holds that empty space exists as an immobile, infinite and penetrable referent in which bodies are located and move from one position to another. The 'Réflexions sur l'espace et le tems' ('Reflections on space and time', hereafter 'Réflexions'), a *mémoire* presented at the Berlin Academy of Sciences in 1748 and published in 1750, constitutes Euler's main contribution to the question. The *mémoire* contains two central arguments based on the law of inertia, the first related to the principle of rest, the second related to the principle of motion and direction. Euler maintains that space must exist independently of our perception and imagination; otherwise, mechanical principles would not be sufficiently grounded. According to Euler's view, the Newtonian metaphysical doctrine, therefore, better accords with the physical laws that are commonly accepted by natural philosophers in the mid-eighteenth century.[1]

1. The 'Réflexions sur l'espace et le tems', in *Histoire de l'Académie royale des sciences et belles lettres* (1748; Berlin, Haude and Spener, 1750), p.324-33, is the only text in which Euler maintains such a strong position. In his other works, he affirms

The same paper contains another argument concerning the representation of space, which has been less discussed in the literature. It is this argument that I would like to focus on here.[2] According to Euler, space cannot be directly perceived by the senses but must be conceived of with the help of reflection. Contrary to the dominant Lockean view, according to which space is perceived by sight and touch as a simple idea, Euler maintains that the contribution of reflection is here necessary. Yet this leads to important problems: what does Euler mean by reflection, and in what way does such a faculty contribute to the representation of space? At the time, there were three main definitions of the capacity of reflection, each of which I will have occasion to discuss here: first, the Lockean view expressed in *An Essay concerning human understanding* (hereafter *Essay*), which holds that reflection is a source of knowledge that originates in our inner perception, and which furnishes us with our ideas about our own cognitive operations; second, Condillac's view, primarily expressed in the *Essay on the origin of human knowledge* (*Essai sur l'origine des connaissances humaines*, hereafter *Essay on the origin of human knowledge*) of 1746, according to which reflection is the ability to focus our attention on the different qualities of an object – in this sense, reflection is not a source of knowledge, but can be associated with what Locke calls discerning;[3] and third, Leibniz's

that there is probably no satisfactory metaphysical explanation for the reality of space, particularly in the *Anleitung zur Naturlehre*, in *Opera omnia*, series 3, vol.1 (Leipzig, 1926), p.16-180 (36-37), in which he has doubts about the possibility of arriving at an appropriate solution. On another less philosophical occasion, that is, in the *Mechanica sive motus scientia analytice exposita* (St Petersburg, Petropoli, 1736), vol.1, def.1, Scol. 1-2, Euler claims that we need not necessarily define the idea of infinite space, and rather that it can be understood as a purely mathematical concept, such that would permit us to leave aside those metaphysical discussions about its true reality. In sum, the doctrine maintained in the 'Réflexions' cannot be considered to be Euler's only answer to the question of the reality of space, even though it is his most explicit contribution to the topic. All these sources will be examined below.

2. For analyses of Euler's principles of rest and motion in the 'Réflexions', see Wallis A. Suchting, 'Euler's reflections on space and time', *Scientia: an international review of scientific synthesis* 104 (1969), p.270-78; Frédéric de Buzon, 'La question de l'espace chez Euler et sa réfraction dans la période précritique: à propos des *Réflexions sur l'espace et le temps* (1748)', *Les Cahiers philosophiques de Strasbourg* 26 (2009), p.141-56.
3. Locke, *An Essay concerning human understanding*, ed. Peter H. Nidditch (Oxford, 1975), II.xi, p.155-63.

view, which argues that reflection is the faculty that enables us to perceive ourselves, and in virtue of which eternal and rational truths are knowable.[4] In this last sense, reflection is related to what one might call self-perception. In this chapter, I wish to determine the following: to which of the three aforementioned definitions of the faculty of reflection does Euler himself most likely subscribe? More precisely, how is reflection for him different from sensation and does this explain, at least partially, the origin of our idea of space? A solution to this problem will certainly help to understand the general argumentative structure of the 'Réflexions' and will shed some light on Euler's position on the status of space.

Space is not the object of sensation

Euler addresses the question of the origin of the idea of space in paragraphs 13 to 15 of the 'Réflexions'. His intention is, in part, to refute Leibniz and his followers, who maintain that space is an imaginary and abstract idea, rather than a real constituent of the world.[5] More precisely, metaphysicians, as Euler calls them, try to reduce space to a phenomenal order of coexistence, in so far as the world is, according to their view, primarily constituted not by spatio-temporally located bodies, but rather by monads or elements endowed with perception and force.[6] On the contrary, Euler aims to show that our idea of space must correspond to a real determination in the world. In this way, space would be as real as corporeal properties,

4. Leibniz, *Monadology*, in *Die philosophischen Schriften*, ed. C. I. Gerhardt, 7 vols (Hildesheim, 1965) (hereafter GP, citing by volume and page number), VI, p.612; *Principes de la nature*, GP VI, p.600-601.
5. As is well known, the main source of the eighteenth-century discussions concerning space and time is the correspondence between Leibniz and Clarke. This correspondence circulated by means of two publications in particular: Samuel Clarke, *A Collection of papers, which passed between the late learned Mr. Leibniz, and Dr. Clarke, in the years 1715 and 1716* (London, James Knapton, 1717); *Recueil de pièces diverses sur la philosophie, la religion et l'histoire par Leibniz, Clarke et Newton*, ed. Pierre Des Maizeaux (Amsterdam, H. Du Sauzet, 1720). Leibniz's last letter is the most detailed and contains the idea that space and time are ideal; see 'Fifth letter of Leibniz to Clarke', GP VII, p.396.
6. See 'Fourth letter of Leibniz to Clarke', GP VII, p.376. Wolff revives the Leibnizian idea in his metaphysical works; see Christian Wolff, *Philosophia prima, sive Ontologia*, in *Gesammelte Werke*, section II, vol.3 (Hildesheim, 2011), §590, p.455-56.

such as inertia, impenetrability and motion. Like Newton's opening definitions and scholium,[7] however, the perception of space is not directly caused by sensation. Euler writes:

> Therefore, metaphysicians are wrong when they want to banish space and place entirely from the world, maintaining that they are only abstract and imaginary ideas. Consequently, the proofs they provide in order to maintain their sentiment, obvious as they may be, must hide some paralogism. It is true that the senses are not capable of giving us the ideas of space and place, and it is only through reflection that we form such ideas.[8]

I will, of course, come back to the last part of this quotation, regarding the reflexive ideas of space and location, but let us now focus on the first epistemological claim. Euler clearly asserts that space is not the object of sensation. On this point, he even appears to subscribe to an anti-empiricist perspective:[9] the idea of space cannot be obtained directly by sight or by touch and would require the intervention of reflection. This statement seems to express Euler's implicit way of distinguishing his own theory from Locke's. In his *Essay*, Locke indeed holds that space is a simple idea given in visual and tactile perceptions. Space is distinct from both the idea of motion and that of body, the latter, for instance, being primarily grasped via the properties of extension and solidity, or resistance. I quote a well-known passage from the *Essay*:

> Motion can neither be, nor be conceived without Space: and yet Motion is not Space, nor Space Motion: Space can exist without it, and they are very distinct *Ideas*; and so, I think, are those of Space and Solidity. Solidity is so inseparable an *Idea* from Body, that upon

7. See Isaac Newton, *The Principia: mathematical principles of natural philosophy*, translated by I. Bernard Cohen and Anne Whitman (Berkeley, CA, 1999), 403-15.
8. Euler, 'Réflexions', §14, p.329. 'Les Métaphysiciens ont donc tort, quand ils veulent bannir entièrement du monde l'espace et le lieu, en soutenant que ce ne sont que des idées abstraites et imaginaires. Par conséquent les preuves qu'ils apportent pour maintenir leur sentiment, quelques fortes qu'elles puissent paraître, seront en effet mal fondées, et il faut qu'il y soit caché quelque paralogisme. Il est vrai que les sens ne sont pas capables de nous fournir les idées de l'espace et du lieu, et que ce n'est que par réflexion, que nous nous formons ces idées.' All translations are my own unless otherwise indicated.
9. On this point, the Newtonian solution relies, as is well known, on the calculus of force: *The Principia,* definitions, scholium, p.410-11.

> that depends its filling of Space, its Contact, Impulse, and Communication of Motion upon Impulse [...] *Space is not Body*, because it includes not the *Idea* of Solidity in it, *Space and Solidity* being *as distinct Ideas.*[10]

Bodies possess the property of solidity that space lacks. Motion is also distinct from space, although no body in motion can be conceived of without space. Thus, space is a simple idea that must be empirically distinguished from body and motion. This also means that we perceive space with the help of our sensory organs, namely sight and touch.[11] The composed ideas of physical objects are constituted by numerous simple ideas, including space. In other words, the idea of space, like solidity and motion, is given directly in sensation.[12]

In the mid-eighteenth century, many other philosophers consider space to be a simple idea obtained first in sensory experience, and therefore share in part Locke's empiricist interpretation. For example, Lambert, who was Euler's colleague for a short time at the Berlin Academy, asserts that space belongs to a limited number of simple concepts composing all of our other notions and principles. One of Lambert's central aims, mainly in the *Anlage zur Architectonic* (*Plan for an architectonic*), is to show that simple concepts are necessary in order to make our analyses of things comprehensible. Locke is even credited here as the one who first achieved an anatomy of concepts, and as the thinker from whom one should take one's inspiration.[13] As Locke claims, space or extension is considered by Lambert as a simple notion at the ground of many scientific realms, including geometry. Without it, not only would it be impossible to assign a clear and primary object to geometry, but so too would it be impossible to determine mathematical quantifications as properties of real and concrete things. Therefore, Lambert defends the thesis according to which space is primarily a notional constituent of the

10. Locke, *Essay*, II.xiii.11, p.172.
11. Locke, *Essay*, II.xiii.2, p.167.
12. Numerous studies have been devoted to this question in Locke; see, for example, Roger Woolhouse, 'Locke's idea of spatial extension', *Journal of the history of philosophy* 8:3 (1970), p.313-17; Geoffrey Gorham and Edward Slowik, 'Locke and Newton on space and time and their sensible measures', in *Newton and empiricism*, ed. Zvi Biener and Eric Schliesser (Oxford, 2014), p.119-37.
13. Johann H. Lambert, *Anlage zur Architectonic*, in *Philosophische Schriften*, vol.3 (Hildesheim, 1965), p.8.

physical world. And most importantly, Lambert maintains, as Locke, that the notion of space is given firstly through sensation.[14]

One reason for which it is possible to imagine that Euler endorses such a conception of spatial sensation is that his theory of knowledge is likewise inspired by the Lockean tradition. Though Euler did not devote many texts to epistemological problems, many of his letters to the princess Frederike Charlotte contain considerations on the source and constitution of our ideas. I will return to some of his thoughts later; here I simply wish to highlight the general conformity of Euler's epistemology with the common empiricist philosophy of the time. For instance, his views on generic and specific notions are very close to Locke's doctrine of nominal essences.[15] The same is true, as we shall see, with the doctrine of abstraction, which is obviously inspired by Locke's *Essay*.[16] However, the 'Réflexions' reveal at least one point of disagreement: for Euler, space, and it also seems to be the case with time, is not given in sensation. This means that we do not directly perceive space by our external senses, as is the case with motion, impenetrability and other corporeal qualities.[17]

According to the *Anleitung zur Naturlehre* (*Guide for the doctrine of nature*),[18] the main problem with the representation of space is that it shares with bodies a sensible property, making it difficult to distinguish it from them: both empty space and bodies have some sort of extension. As previously mentioned, Locke gives an answer to this problem: solidity must be added to extension in order to conceive of the idea of a body, while space lacks solidity or resistance. This is problematic, since space could not be directly given with the help of a sensible perception but would rather have to be discerned with

14. Lambert, *Anlage zur Architectonic*, p.51. The empirical origin of simple concepts is, however, completed by *a priori* reasoning in Lambert's doctrine. Space is not only given in sensation but also theoretically presupposed in any division of its parts. See Lambert, *Neues Organon*, 'Alethiologie', in *Philosophische Schriften*, vol.1 (Hildesheim, 1965), §12 and 52, p.458-59 and 487.
15. Leonhard Euler, *Lettres à une princesse d'Allemagne sur divers sujets de physique et de philosophie*, in *Opera omnia*, series 3, vol.11-12 (Basel, 1960), vol.12, §101.
16. Euler, *Lettres à une princesse*, vol.12, §100.
17. Euler makes this assertion in the 'Réflexions', despite his general approval of empiricism expressed in the *Lettres*, according to which sensible ideas are at the foundation of all other knowledge: 'et c'est par le moyen des sens que l'âme parvient à la connaissance de ce chien et en général des objets externes, et qu'elle en acquiert les idées sensibles, qui renferment le fondement de toutes nos connaissances' (Euler, *Lettres à une princesse*, vol.12, §98).
18. Euler, *Anleitung*, p.16-178.

the help of a fairly complex analytical procedure. On the one hand, bodies would be directly perceived, since extended and impenetrable material parts can easily be observed in the realm of our empirical knowledge. On the other hand, space seems to be necessary in order to have a complete representation of the physical world, even though there appears to be no way to associate space with a single sensible property. Here is what Euler asserts in the *Anleitung zur Naturlehre*:

> Everything that is impenetrable belongs to the corporeal kind, and thus the essence of bodies consists in their impenetrability, in which consequently all other properties must find their reason. The essence of material bodies consists in such a property, which is not only common to all bodies, but is also so peculiar that it does not belong to any other thing. For this reason, the essence of bodies cannot be located in extension, because space is also extended.[19]

In the *Anleitung zur Naturlehre*, Euler does not give a proper demonstration of the existence of space, as he does in the 'Réflexions'. But the problem seems to be more or less the same: it is impossible to perceive space in sensation, for there is no attribute or property obtained with the help of the senses that belongs exclusively to this entity, and which would therefore enable one to express its proper nature. This appears to be the main reason why Euler claims that the mind does not perceive the idea of empty space via sensation. There is no distinct access to this entity through sensory perception, since extension is common to both bodies and space. The mind must therefore separate space from our representation of physical objects, which would require something more than sensation alone.

Space is not a generic idea

In order then to conceive of a genuine representation of empty space, Euler turns to reflection. As we have seen, the 'Réflexions' are quite clear on this point: in order to perceive space, the mind must exercise its reflexive capacity, not its sensory one. As was already mentioned,

19. Euler, *Anleitung*, p.49: 'Alles was undurchdringlich ist, gehört in das Geschlecht der Körper, und daher besteht das Wesen der Körper in der Undurchdringlichkeit, in welcher folglich alle übrigen Eigenschaften ihren Grund haben müssen. Das Wesen der Körper besteht in einer solchen Eigenschaft, welche nicht nur allen Körpern gemein, sondern auch so eigen ist, dass sie keinen andern in der Ausdehnung gesetzt werden, weil der Raum auch ausgedehnt ist.'

the three main definitions of reflection which persist during this period are advanced respectively by Locke, Condillac and Leibniz. First, it seems evident that, for Euler, reflection does not mean, at least in this context, the perception of the operations of one's own understanding, as Locke defines it in the *Essay*. Locke makes a distinction between sensation and reflection, claiming that the former is a source of learning from the external senses, while the latter provides the knowledge of one's own cognitive capacities (i.e. perceiving, judging, willing, etc.,) by means of an inner perception.[20] Obviously, Euler does not take reflection in this sense, that is, in terms of the inner perception of our own faculties, when he states that the idea of space is obtained through reflection. At least in the 'Réflexions', space is not considered as a mere cognitive or imaginary content; it would even appear to be the opposite, for space is a constituent of the physical and external world. Thus, it is not reasonable to imagine that Euler could have revived Locke's conception of reflection in his characterisation of our knowledge of space.

Perhaps Euler might have in mind Condillac's definition. In the *Essay on the origin of human knowledge*, Condillac devotes an entire chapter to reflection. For him, as is the case for Locke, all ideas and principles are derived from experience, and thus none is innate. However, contrary to Locke, Condillac does not claim that some of our ideas are acquired via inner perceptions: all ideas are taken from sensation, even those that characterise our cognitive faculties. Ideas representing the mind and its capacities are as sensible as those corresponding to external and bodily phenomena.[21] This means that reflection is not a source of knowledge, as it is for Locke, but a cognitive power working from sensible ideas. More precisely, Condillac associates reflection with attention, or with what Locke would have called discernment.[22] Discernment refers to one's ability to distinguish or separate ideas from each other: 'This manner of applying, on our own, attention to different objects in succession, and to different parts of a single object, is called "to reflect". Thus, we clearly see how reflection is born from imagination and memory. But there is a progression that we must understand.'[23]

20. Locke, *Essay*, II.i.4, p.105.
21. Etienne Bonnot de Condillac, *Essay on the origin of human knowledge*, translated by Hans Aarsleff (Cambridge, 2001), p.9.
22. Condillac, *Essay on the origin of human knowledge*, p.155-63.
23. Condillac, *Essay on the origin of human knowledge*, p.41.

In this way, reflection is based primarily on the operations of the imagination and memory, which yield sensible contents, and thus allow one to separate different qualities of the represented objects or properties. In other words, reflection is the capacity of dividing complex sensible contents into simpler ideas. The mind's attention can break down any complex idea as long as it perceives different properties or qualities in it. Once again, reflection is not, for Condillac, a source of knowledge, but a cognitive power for discerning basic notions from more composite ones.

One central role played by reflection is that it separates qualities from a particular context of perception and thereby establishes abstract or general ideas. Not only can one's attention focus on a plurality of distinct qualities, but it also has the power to distinguish qualities that are common to several objects.[24] This is another difference between Locke and Condillac: while Locke seems to maintain that abstraction can be made from both sensible and reflexive ideas,[25] Condillac claims that it is the function of reflection alone to conceive of general and abstract notions. Since reflection is understood as an attentive faculty, it is the only cognitive power aiming at discerning properties from their particular sensible context. An abstract idea is the result of the reflexive perception and is not linked to the other properties of a particular object.

This is very interesting for us, since Condillac explains, later on in his *Essay on the origin of human knowledge*, that the origin of our perception of space is grasped precisely with the help of such an abstraction: the idea of space results from a double abstraction made possible by reflection. We first separate the idea of the extended and moveable body from our perception of particular sensible objects, and then proceed to a second abstraction from this idea of body, in order then to obtain the notion of pure space:

> If we cease to think of the substance of bodies as being actually coloured and figured, looking on it only as something mobile, divisible, impenetrable, and having an indeterminate extension, we shall have the idea of matter, an idea simpler than that of body, of which it is only an abstraction, though many philosophers may have had the fancy to reify it. If afterwards we cease to think of the mobility of matter, its divisibility and impenetrability, in order to

24. Condillac, *Essay on the origin of human knowledge*, p.24.
25. Locke, *Essay*, II.xi.9, p.159.

reflect only on its indeterminate extension, we will form the idea of pure space, which is still simpler.[26]

Two aspects must be underlined here: on the one hand, reflection reaches simpler abstract ideas by discerning or separating them from complex perceptions. The idea of matter is more basic than the ideas of particular objects. In the same way, pure space is simpler than the idea of a moveable and impenetrable material body, because it is abstracted from a more complex representation. Thus, Condillac rehabilitates a fairly classical model, one that can be found in Locke's philosophy, as well as before it: here, the concept of analysis is defined as breaking down one's empirical complex representations into simple ideas. On the other hand, this analytical operation fits into a structure of mereological dependency. The faculty of reflection enables us to differentiate ideas, but also to compare their conceptual relationships and form general archetypes or models.[27] In this way, the power of reflection seems to result in ideas of kinds and species that express general similarities between objects. The idea of matter is thus the common species abstracted from individual and corporeal objects, whereas the idea of pure space would be a kind common both to bodies and to particular objects. The more general ideas are less complex than they seem to be. Of course, for Condillac, these abstract ideas are ways of conceiving of things, and not real beings. This is why the idea of space would be a generic and purely mental content, even though it is a constituent of our representation of the world. The idea of space is thus obtained through the faculty of reflection, which is capable of abstracting or separating properties from more specific or individual ideas.

Let us go back to our main discussion. The question is whether Condillac could have influenced Euler's understanding of reflection. Some contextual evidence might support this claim. First, the *Essay*

26. Condillac, *Essay on the origin of human knowledge*, p.205. 'Si nous cessons de penser à la substance des corps comme étant actuellement colorée et figurée, et que nous ne la considérions que comme quelque chose de mobile, de divisible, d'impénétrable et d'une étendue indéterminée, nous aurons l'idée de la matière: idée plus simple que celle des corps, dont elle n'est qu'une abstraction, quoiqu'il ait plu à bien des philosophes de la réaliser. Si ensuite nous cessons de penser à la mobilité de la matière, à sa divisibilité et à son impénétrabilité, pour ne réfléchir que sur son étendue indéterminée, nous nous formerons l'idée de l'espace pur, laquelle est encore plus simple'; *Essai sur l'origine des connaissances humaines*, in *Œuvres philosophiques de Condillac* (1746), ed. Georges Leroy, vol.1 (Paris, 1947), p.109.

27. See Condillac, *Essay on the origin of human knowledge*, p.24.

on the origin of human knowledge was published in 1746, two years before Euler read the 'Réflexions' at the Berlin Academy. Condillac's work spread quickly, not only in France, but also throughout other European countries, including Germany. Euler and his colleagues in Berlin, in particular Maupertuis and Formey, most likely discussed the content of the *Essay* shortly after its publication. Besides, Condillac was elected as a corresponding member to the Berlin Academy in 1749, which shows that his work gained rapid recognition at Frederick II's court. Also, when we take a closer look at Euler's general epistemology, primarily represented in the *Lettres à une princesse d'Allemagne* (*Letters to a German princess*), we immediately see the theoretical relationship between his doctrine and Condillac's conception of ideas. This is particularly true of his theory of abstraction, which is grounded in his concept of attention. According to Euler, attention enables us not only to discern properties, but also to arrive at more general ideas. The similarity between Condillac's and Euler's views, both originally inspired by Locke, is quite obvious:

> Here the soul exerts a new faculty, which is called the power of abstraction; this takes place when the soul fixes its attention on only one quantity or quality of the object, and considers it separately, as if it were no longer attached to the object. [...] These ideas, acquired by abstraction, are denominated notions, to distinguish them from sensible ideas, which represent to us really existing objects.[28]

Like Condillac, Euler believes that abstraction is made possible with the help of attention. He gives the example of the general idea of heat, which can be found in many objects, such as stone, wood and water, but is not reducible to these individual and sensible realities. General notions are not attached to sensible objects and thus can be expressed independently of them. Moreover, abstract ideas permit one to know the essential properties of things. Contrary to animals, the human being is not limited to perceiving sensible and individual things, but can express their general essences.[29] Attention endows us

28. Euler, *Lettres à une princesse*, vol.12, §100: 'C'est ici que l'âme déploie une nouvelle faculté, qu'on nomme l'abstraction, qui se fait, quand l'âme fixe son attention uniquement sur une quantité ou qualité de l'objet, qu'elle l'en sépare et la considère comme si elle n'était plus attachée à l'objet. [...] Ces idées acquises par l'abstraction sont nommées notions pour les distinguer des idées sensibles, qui nous représentent des choses réellement existantes.'
29. Euler, *Lettres à une princesse*, vol.12, §109.

with the ability to represent to ourselves abstract objects, and thereby makes possible knowledge of what Locke calls ideas of species, or nominal essences.[30]

For both Condillac and Euler, attention is the capacity for separating out ideas, and makes possible the conception of essences. For his part, Condillac has defined reflection as the application of attention to the sensible contents of imagination and memory. Moreover, he thinks that reflection enables us to perceive the notion of pure space, since it requires abstract procedures. Therefore, there seem to be some good reasons for maintaining that Euler also agrees on this point. Reflection, taken in Condillac's sense, might be the faculty that provides us with an abstract idea of space. However, two elements prevent us from considering this interpretation valid. First, Euler never clearly claims that reflection is reducible to some mental capacity for focusing on simpler and more abstract properties. In the *Lettres à une princesse*, there is no strong association between reflection and attention. So, even if Euler had read Condillac's *Essay on the origin of human knowledge*, which my analyses above seem to suggest, one finds no evidence in his work – whether in the 'Réflexions', or in the *Anleitung zur Naturlehre* or the *Lettres à une princesse* – that would suggest Euler had incorporated these modes of attentiveness and abstraction into his understanding of the faculty of reflection. Euler might, therefore, have conceived of these abilities as independent of the faculty of reflection. Secondly, and most importantly, Euler rejects the explanation that space is apprehended by means of abstraction, at least he does so in the 'Réflexions'. Of course, as previously noted, we arrive at many general ideas with the help of attention and abstraction. Such is the case, for instance, with respect to our understanding of heat and extension as qualities separable from particular bodies. But the concept of space is not based on such an abstraction. Here is a passage from the 'Réflexions':

> I agree that all the things that exist are perfectly determined, and if we subtract one or a few determinations from the idea of such an object, a generic idea arises from this operation, and to which does not correspond any existing object. This is how we form the idea of extension in general, by subtracting all determinations from ideas of bodies, apart from extension. But the idea of the place that a body occupies is not formed by subtracting some determinations of the

30. Locke, *Essay*, III.vi.11, p.445-46.

> body; it results from removing the whole body, so that the place cannot be a determination of the body, since it still remains after removing the whole body with all its quantities. For it must be noted that the place that a body occupies is quite different from its extension because extension belongs to the body, and comes with it through the motion from one place to another, while place and space are not susceptible to any motion.[31]

In the last section of the chapter, I wish to return to several ideas expressed in this quotation, but for now let me draw attention to one particular point. In general, Euler seems to accept Condillac's position, which holds that abstraction grasps generic or specific notions, that is, essential properties of things. Extension, motion and impenetrability are abstract ideas that express the essence of bodies. On this point, there seems to be no major disagreement between Euler and Condillac, for both theories originate from Locke's view. Knowledge of essential properties is derived from the capacity for abstracting. However, it appears that Euler does consider at least one exception, which pertains to the idea of space. As indicated in the former passage, space and location are not obtained in the same way as extension and motion: we do not grasp the idea of space by abstracting it from the properties of corporeal things, for it is not a corporeal property. Instead, acquiring the idea of empty space requires us to subtract the entire body. Obviously, space is not a generic quality of bodies, but is completely distinguishable from them. Although Euler recognises that some sort of cognitive separation is necessary in order to conceive of empty space, it has little to do with the act of abstraction that one finds in Condillac. According to Euler, we do not abstract space from bodies, as if it were a generic property

31. Euler, 'Réflexions', p.329-30: 'Je suis d'accord que toutes idées qui existent, sont parfaitement déterminées; et, si nous retranchons de l'idée d'un tel objet, une ou quelques déterminations, qu'il en naît une idée générique, à laquelle il ne répond plus d'objet existant. C'est ainsi que nous nous formons l'idée de l'étenduë en général, en retranchant des idées des corps toutes les déterminations, hormis l'étenduë. Mais l'idée du lieu qu'un corps occupe, ne se forme pas en retranchant quelques déterminations du corps; elle résulte en ôtant le corps tout entier: de sorte que le lieu n'ait pas été une détermination du corps, puisqu'il reste encore, après avoir enlevé le corps tout entier avec toutes ses quantités. Car il faut remarquer, que le lieu qu'un corps occupe est bien différent de son étenduë, parce que l'étendue appartient au corps, et passe avec lui par le mouvement d'un lieu à l'autre; au lieu que le lieu et l'espace ne sont susceptibles d'aucun mouvement.'

of corporeality, rather we must separate space from any bodily determinations, such as impenetrability or motion, because space is precisely what is penetrable and immoveable. Thus, it is impossible that Euler could have understood reflection as some faculty in virtue of which one abstracts space as if it were a general kind or species. By reflection, he cannot mean here a capacity of abstracting simpler and more general ideas from particular and individual sensible perceptions. Certainly, this kind of cognitive operation is possible, and even necessary to the expression of the properties of corporeal things or sensible qualities. However, the representation of space is a particular case that requires another kind of mental act, which is different from the operation of abstracting generic and specific properties from individual objects.

Body and space

What exactly is this cognitive procedure on which the perception of space is based? The *Anleitung zur Naturlehre* probably contains the most detailed argumentation in this regard and therefore deserves closer examination. The first important element is that space seems representable only when it is compared to bodies. The relationship with the determination of corporeal things would thus be the basis for the idea of empty space:

> From any concept of space that we make, we cannot conceive of any other body than the one occupying a certain part of space, and so to speak filling it. This part of space occupied by a body, which must be the same as the extension of the body, is called its location; even if a body cannot be at more than one location at the same time, it is however not linked to it, so that it could not at another time be at another location. For neither in space, nor in the body itself is there anything that this possibility would be contradicted. The possibility that a body is transferred from one location to another is so inherent to our concept of bodies, that we cannot perceive any body for which this should not be possible.[32]

32. Euler, *Anleitung*, p.29: 'Was wir uns auch für einen Begriff von dem Raume machen, so können wir uns keinen Körper anders vorstellen, als dass er einen gewissen Teil des Raumes einnimmt und gleichsam ausfüllt. Dieser von einem Körper eingenommen Teil des Raumes, welcher der Ausdehnung nach dem Körper vollkommen gleich sein muss, wird sein Ort genannt; ob nun wohl ein Körper zugleich an nicht mehr als einem Orte sein kann, so ist er doch nicht dergestalt daran verknüpfet, dass er nicht zu einer anderen Zeit sich an einem

The main argument pertains to the conceptual relationship between the properties of bodies and space. The very notion of a body in motion or at rest would not be conceivable if it were not accompanied by the idea of an empty space in which it is located, or in virtue of which it could move from one place to the other. The reasoning here is very close to what Locke maintains in the *Essay*, as well as that upon which I have previously insisted.[33] Again, empty space is not directly perceived, as corporeal and sensible qualities are, but is nonetheless involved in the perception of the physical world. Even if a body is at rest, it is conceived of as occupying a position, a reality that is different from the body itself. The same is true of any perception of motion that involves a shift of the body from one place to another. Thus, it is clear that a complete representation of the external world entails the idea of an immoveable referent.

Therefore, the comparison with corporeal beings is required to determine the characteristics of space. Apart from being immoveable, Euler recognises two other main properties of space. First, space is infinite.[34] Here, Euler agrees with Newton, who considered absolute space to be an infinite form of the universe.[35] This is clearly another distinction between space and body: the latter must have boundaries, while the former is without such limitations and is therefore infinite. Since bodies are limited, we can infer that space, in which such limited bodies are situated, is, inversely, unlimited and infinite. Secondly, space, in contrast to bodies, is penetrable. In many texts, Euler insists on saying that impenetrability is one of the main properties – along with extension, motion and inertia – of corporeal things.[36] More precisely, impenetrability seems to presuppose space: the idea that a body cannot penetrate another body is grasped with the help of the notion of location, since being impenetrable also means that two things cannot occupy the same place in space.[37] Thus, Euler attributes penetrability to space in more or less the same way as he does immobility and infinity, that is, in comparison with bodily determinations. Since corporeal beings are located in space or

---

anderen Orte sollte befinden können; denn weder in dem Raume noch dem Körper selbst ist nichts, wodurch dieser Möglichkeit widersprochen würde.'

33. Locke, *Essay*, II.xiii.11, p.172.
34. Euler, *Anleitung*, p.36.
35. Newton, *The Principia*, 'General scholium', p.941.
36. Euler, *Anleitung*, p.45-50; *Lettres à une princesse*, vol.12, §119-20.
37. Euler, *Anleitung*, p.45-46.

can change their *situs*, this means that space must be penetrable. In sum, space is immobile, infinite and penetrable, for it represents the different positions in which mobile, finite and impenetrable bodies are situated. Euler even claims that the idea of body could be understood in two different ways, externally or internally, depending on whether one is referring to the body's inner constitution or to its relation to empty space:

> We can perceive a double state in a body, the external one and the internal one. The latter consists in the kind of parts, which constitute the body, and its composition itself; but the external state, of which alone we are speaking here, consists in the relations of the body with space. As long as a body is at rest, it remains in the same location and there is no doubt that it must persist in the same relation to space. When a body moves, it continually changes its location, but when this occurs always in the same direction and at the same speed, then there remains in the change of location a constant sameness, so that also in this case one can say that the relation to space remains the same. In both cases one says that the body persists in the same state.[38]

Here one notices that space is necessary in order to have an adequate representation of the external state of bodies. It seems evident that the idea of space must be conceived of by analogy with that of an inner state. Space is perceived by comparison with the inner characteristics of bodies. This is consistent with what Euler claims in the 'Réflexions', in which he maintains that space is not the immediate object of sensation. In order to grasp the idea of space, a conceptual analysis of bodily properties, which are perceived by sensible experience, is required. More specifically, the idea of space

38. Euler, *Anleitung*, p.41: 'Man kann sich in einem Körper einen doppelten Zustand vorstellen, den äußerlichen und den innerlichen. Dieser besteht in der Art der Teile, aus welchen der Körper bestehet, und ihrer Zusammensetzung selbst; der äußerliche Zustand aber, von welchem all hier allein die Rede ist, besteht in den Verhältnissen des Körpers mit dem Raume. So lange sich nun ein Körper in Ruhe befindet, so bleibt er an ebendemselben Ort und ist nun ein Körper in Ruhe befindet, so bleibt er an ebendemselben Ort und ist also kein Zweifel, dass er nicht in ebendemselben Verhältnisse mit dem Raume verharren sollte. Wenn sich aber ein Körper beweget, so verändert er zwar beständig seinen Ort; wenn aber dieses immer nach einerlei Richtung und mit ebenderselben Geschwindigkeit geschieht, so bleibt in der Veränderung des Orts selbst eine beständige Gleichheit und daher lässt sich auch in diesem Fall sagen, dass Verhältnis gegen den Raum einerlei bleibe.'

is attained when all bodies are subtracted from it. In other words, our representation of space is indirectly inferred through theoretical oppositions along with the corporeal and sensible properties, from which they must be distinguished. The external state of bodies, which expresses its relation to empty space, is contrasted with the internal one, which consists of the corporeal properties of impenetrability, motion and inertia that represent the essence of bodies. This kind of argument exceeds sensible perception and is based on the cognitive operations of comparison and separation. An inverse analogy is necessary, beginning with the general sensible properties, those of bodies, and arriving at non-sensible or reflexive ones – immobility, penetrability and infinity. We must also recall that the idea of space is obtained not by abstracting a generic essence from bodies, but rather by removing the entire body and thereby attaining a representation of empty space with the help of reflection.

A Leibnizian influence?

Does the use of reflection only signify a kind of argument in which the properties of space are inferred by comparison with corporeal beings? Would that mean that Euler does not refer to a more precise definition of this faculty? I would like to complete this analysis by putting forward one last hypothesis. In my opinion, there seems to be a link between Euler's and Leibniz's conceptions of reflection (i.e. the third view mentioned earlier). This might seem a surprising interpretation, since, as previously stated, Euler heavily criticises Leibniz's philosophy, as well as many of his followers, especially Wolff. In both the 'Réflexions' and the *Lettres à une princesse*, Euler opposes many Leibnizian doctrines, for example, the theory of monads and the doctrine of pre-established harmony.[39] As is also well known, during the years following the publication of the 'Réflexions', Euler took Maupertuis's side in the controversy opposing him to Samuel König, who maintained that the Principle of least action was already present in Leibniz.[40] Nevertheless, it seems very plausible to argue that Euler defines the faculty of reflection in a Leibnizian way. Considering

39. For a more detailed analysis, see Christian Leduc, 'Euler et le monadisme', *Studia Leibnitiana* 45:2 (2013), p.150-69.
40. For historical and theoretical details on the Principle of least action, especially on Euler's contribution, see Alberto Rojo and Anthony Bloch, *The Principle of least action, history and physics* (Cambridge, 2018).

that space is obtained neither through sensation nor through generic abstraction, as Locke and Condillac respectively maintain, I am inclined to think that Euler's concept of reflection is similar to that of Leibniz.

Here is the evidence in support of this claim. First, Euler was quite familiar with a number of Leibniz's works, and had without a doubt read his *Essays on theodicy* (*Essais de théodicée*), the *Monadology* (*Monadologie*), as well as his correspondence with Samuel Clarke. Second, his assertions in the 'Réflexions' regarding the knowledge of space seem to accord with Leibniz's own concept of the knowledge of space; it goes without saying that such assertions are made independently of Leibniz's views on the ontological status of space, to which Euler is strongly opposed.

What exactly is Leibniz's definition of reflection? Significant research has been conducted on this aspect of Leibniz's philosophy; here, however, we might permit ourselves to focus only on those aspects of his thought that are necessary for the sake of the present discussion.[41] These are raised in the *Monadology*:

> It is also through the knowledge of necessary truths and through their abstractions that we rise to reflective acts, which enable us to think of that which is called the 'I' and enable us to consider that this or that is in us. And thus, in thinking of ourselves, we think of being, of substance, of the simple and of the composite, of the immaterial and of God himself, by conceiving that that which is limited in us is limitless in him. And these reflective acts furnish the principal objects of our reasonings.[42]

Of course, Euler would not accept the notion that all of these ideas could somehow be perceived by an act of reflection. He remains on

41. See especially Robert McRae, *Leibniz: perception, apperception, and thought* (Toronto, 1976); Mark Kulstad, *Leibniz on apperception, consciousness and reflection* (Munich, 1991).
42. Gottfried Wilhelm Leibniz, *Monadology*, in *G. W. Leibniz: philosophical essays*, ed. and translated by Roger Ariew and Daniel Garber (Indianapolis, IN, 1989), p.213-25 (217); GP VI, p.612: 'C'est aussi par la connaissance des vérités nécessaires et par leurs abstractions que nous sommes élevés aux actes réflexifs, qui nous font penser à ce qui s'appelle moi, et à considérer que ceci ou cela est en nous: et c'est ainsi qu'en pensant à nous, nous pensons à l'Etre, à la Substance, au simple et au composé, l'immatériel et à Dieu même; en concevant que ce qui est borné en nous, est en lui sans bornes. Et ces actes réflexifs fournissent les objets principaux de nos raisonnements.'

many points very close to Locke's epistemology. However, what must be noticed is the mind's capacity to access, with the help of reflexive acts, conceptual contents, such as the ideas of substance, simplicity and so on. In fact, Leibniz makes two important points with regard to the present topic: first, only some beings that he calls spirits are capable of self-perception or reflection. This faculty is what makes the distinction between human beings and animals, since the former are conscious of their acts of thinking, while the latter are incapable of such knowledge.[43] Animals are surely conscious of their sensory and imaginary experiences, but they will never be able to perform reflective cognitive acts.[44] By contrast, human beings possess such a cognitive power. In this sense, reflection is clearly the capacity of perceiving oneself as a thinking being, that is, as a self. Inner perception makes conception of the self possible, and, therefore, the conception of one's own conceptual activity. This understanding of self-perception is close to the Cartesian definition, but Leibniz is probably the first modern philosopher to link self-perception with reflexivity in this way.[45] Secondly, acts of reflection allow us to conceive of a great number of ideas. As is well known, Leibniz considers such ideas to be innate, just as he maintains that each monad already contains within itself all of its past, present and future perceptions.[46] We can perceive the ideas of substance, simplicity, God and so on because they are present in us from the beginning of our existence. But what I would like to emphasise is that general and necessary truths are obtained not through sensation, but through reflection alone.[47] On the one hand, Leibniz agrees that one's ideas of colour, sound, form and motion are perceived in sensations of them. On the cognitive level, Leibniz believes that sensory experience makes possible the representation of these ideas; without it, the mind

43. Leibniz, *Principes de la nature et de la grâce*, GP VI, p.600.
44. Leibniz, *Principes de la nature*, GP VI, p.601; *Monadology*, GP VI, p.611.
45. On this point, see Christian Barth, 'Leibnizian conscientia and its Cartesian roots', *Studia Leibnitiana* 43:2 (2011), p.216-36. Subsequently, Christian Wolff will retrieve aspects of Leibniz's definition; see Wolff, *Psychologia empirica*, in *Gesammelte Werke*, section II, vol.5, ed. Jean Ecole *et al.* (Hildesheim, 1968), p.187.
46. This notion was mostly developed in the *Nouveaux essais sur l'entendement humain*, in *Œuvres philosophiques de feu Mr de Leibniz*, ed. Rudolph Eric Raspe (Amsterdam and Leipzig, Schreuder, 1765), a contribution that Euler could not have read when he wrote the 'Réflexions', as it was only published in 1765.
47. Leibniz, *Monadology*, GP VI, p.611.

would be incapable of representing such sensible contents.[48] On the other hand, Leibniz identifies some notions and truths that are accessible on the basis of reflexive inferences. Eternal truths belong to this second type and are obtained with the help of reflection. Thus, for Leibniz, we can find the origin of many concepts and principles via reflection, which enables us to conceive of things from an intellectual and general representation of them.

As previously raised, this model of cognition is very different from the one that Euler puts forward in both the 'Réflexions' and the *Lettres à une princesse*; Euler offers an empiricist view in most aspects of his theory of knowledge. Besides, Leibniz does not believe that the origin of our representation of space, defined as an order of coexistence, is purely reflexive; the notion of space is obtained by means of imagination, which therefore explains why he considers spatiality to be an ideal determination.[49] Euler and Leibniz would nonetheless agree on the fact that Locke's and Condillac's positions are indefensible. Concerning the problem of the representation of space, Euler seems to have realised that a mere empiricist account leads to major problems. Since space is not directly given in sensation, but inferred by comparison with corporeal properties, it means that the contribution of the understanding, what Euler calls reflection, is required. On this precise point, Euler would thus be in agreement with the Leibnizian view. The previously described operation of comparison between corporeal properties and space resembles Leibniz's thesis according to which some kind of contents are the result of a reflexive inference.

Three main elements support this interpretation: first, Euler agrees with Leibniz that space is not the result of a generic abstraction. He believes that the empiricist thinkers, like Condillac, are wrong in this respect. According to Euler, Leibniz might be right when he claims that space is not deduced from corporeal and sensible properties, even though he is mistaken when he infers that space and location are imaginary.[50] In other words, Leibniz's theory of cognition seems to be partially correct; though his ontology and physics would remain highly problematic. Besides, Euler's main criticisms of Leibniz, which

48. Leibniz, *Meditationes de cognitione, veritate, et ideis*, in *Sämtliche Schriften und Briefe*, ed. Deutsche Akademie der Wissenschaften (Darmstadt, 1923-), VI, 4, p.586.
49. See Leibniz, 'Fifth letter to Clarke', GP VII, p.395-96.
50. Euler, *Réflexions*, p.325.

we essentially find in the *Lettres à une princesse* and the *Gedancken von den Elementen der Cörper* (1746),[51] concern the doctrine of monads and its consequences, not his theory of knowledge. Euler could have been influenced by different traditions, such that they would give rise to a hybrid conception. Although most ideas are taken from sensation, such as those of extension, motion, colours and sounds, certain others could be derived from reflective reasoning, notably the idea of space.

The second reason supporting this interpretation pertains to Euler's very own assertions. Not only does he say that space is conceived of in reflection, but also that we must think about ourselves (*réfléchir à soi-même*) in order to understand that space is not the object of an abstractive procedure based on sensible and corporeal ideas. Here is the complete passage from the 'Réflexions':

> It is true that the senses are not capable of giving us the ideas of space and place, and it is only through reflection that we form such ideas. From this, they [metaphysicians] conclude that these are merely abstract ideas, similar to the ideas of kinds and species, which only exist in our understanding and to which no real object corresponds. But this conclusion seems to be precipitated; for if we think about ourselves, we will easily perceive that the way in which we arrive at the ideas of space and place is quite different from the way in which we form the ideas of kinds and species. And it would be a grave mistake if we were to assert the idea that there is nothing else of which we have no ideas through reflection.[52]

The first part explains what has been previously analysed concerning the reflexive origin of space. However, in the second part, Euler clearly associates our reflexive capacity with that of self-perception, or self-conception, a cognitive operation that enables

51. Leonhard Euler, *Gedancken von den Elementen der Cörper* (Berlin, Haude and Spener, 1746).
52. Euler, 'Réflexions', p.329: 'Il est vrai que les sens ne sont pas capables de nous fournir les idées de l'espace et du lieu, et que ce n'est que par réflexion, que nous nous formons ces idées. De là ils concluënt que ce ne sont que des idées abstraites, semblables aux idées des genres et des espèces, qui n'existent que dans notre entendement, et auxquelles il ne répond aucun objet réel. Mais il me semble que cette conclusion est précipitée: car pour peu qu'on réfléchisse à soi même, on s'apercevra aisément, que la manière, dont on parvient à l'idée de l'espace et du lieu, est bien différente de celle, dont nous nous formons les idées des genres et des espèces. Et on se tromperait fort, si l'on voulait soutenir, qu'il n'existe pas des choses, dont nous n'avons d'autres idées que par réflexion.'

us to attain the representation of space. Of course, he does not say that space and location are innate and intellectual contents. But it seems evident that the comparative argumentation takes place within reflection, which allows us to express concepts in self-perception. Thus, the representation of space would require some sort of intellectual and reflexive inference, which is needed to complete our sensible knowledge of physical properties. Inferring the existence of empty space is based on a reflexive act, similar to those mentioned in Leibniz's *Monadology*; otherwise, no idea of space and location could be given. The association between self-perception and reflection might result in different conceptual contents for both Leibniz and Euler, but they appear to be in general agreement about the very definition of this cognitive capacity.

Finally, one finds here contextual evidence that could reinforce such an interpretation: from the early 1740s, as previously mentioned, Euler was an active and important member of the Berlin Academy. Along with Maupertuis, who was president of the Academy from 1746 to 1759, Euler attempted to convince his colleagues of the merits of Newtonian physics; of course, at the time, most of them subscribed to the predominant Leibnizian and Wolffian doctrines then found in Germany. By this time, opposition to the Leibnizians had become palpable, the affair surrounding the Principle of least action and Samuel König's attempt to credit Leibniz with its formulation being the most well-known examples. But the Berlin Academy also hosted notable Leibnizians and Wolffians, in particular, Jean Henri Samuel Formey, who was the secretary for many years and Johann Georg Sulzer who became a member in 1750. The Academy was therefore a place where at least two essential movements, the Newtonian and the Leibnizian, confronted one another. Many of Euler's colleagues were supporters of Leibniz's theories and could have, in some ways, influenced Euler's epistemological positions. Even more striking is the interest in the question of reflection and self-knowledge at the end of the 1740s and during the 1750s. One year after Euler presented the 'Réflexions', Jean Bernard Merian presented two papers on apperception and reflection.[53] Later, Sulzer produced a *mémoire* on

53. Jean Bernard Merian, 'Mémoire sur l'apperception de sa propre existence', in *Histoire de l'Académie royale des sciences et belles lettres* (1749; Berlin, Haude and Spener, 1751), p.416-41, and 'Mémoire sur l'apperception considérée relativement aux idées', in *Histoire de l'Académie royale des sciences et belles lettres* (1749; Berlin, Haude and Spener, 1751), p.442-77.

the same topic.[54] In all of these works, Leibniz's views are discussed in order to determine the role that reflection and apperception play in the origination and perception of our ideas. This means that Euler was participating in a broader discussion on intellectual and sensible cognitive capacities, at the heart of which was Leibniz's contribution. His position in the 1748 paper must be understood in a context that is, therefore, in part favourable to the Leibnizian epistemology. Perhaps Euler was convinced of the Leibnizian definition of reflection and used it in his own article, even though he generally disagreed with Leibniz's scientific and metaphysical views.

In conclusion, I would like to insist on the discrepancies between Euler's different contributions to the question. As I showed above, the 'Réflexions' put forward a thoroughly Newtonian conception of space by relying on the faculty of reflection. Reflection is what allows us to conceive of space, in so far as it gives us the ability to separate all sensible and corporeal properties of physical objects, and therefore to understand space as an immobile, infinite and penetrable entity. This inference seems possible only through the intervention of a non-sensible faculty. However, the 'Réflexions' is peculiar in two respects: first, it is the only text in which Euler mentions this faculty. In neither the *Mechanica sive motus scientia analytice exposita*, the *Anleitung zur Naturlehre*, nor the *Lettres à une princesse* does he characterise reflection as the capacity for conceiving of non-sensible contents, such as the idea of space. This argument is unique and is not found, as far as I know, in any other texts. Second, these other contributions, in particular the *Anleitung zur Naturlehre*, remain cautious about giving any definitive answer to the question of the reality of space. In these other works, Euler does not really commit himself on the question of its reality. I would like to suggest a reason why this might be the case: Euler was probably not satisfied with his own solution to the problem. Perhaps he realised that both the ontological and the epistemological aspects of the question raised many problems that he was not able to solve. For example, how exactly is it possible for the mind to conceive

54. Sulzer, 'Sur l'apperception et son influence sur nos jugemens', in *Histoire de l'Académie royale des sciences et belles lettres* (1764; Berlin, Haude and Spener, 1766), p.415-34. This text was most probably read in the 1750s, although published several years later. On this point, see Daniel Dumouchel, 'Johann Georg Sulzer: la psychologie comme "physique de l'âme"', in *Les Métaphysiques des Lumières*, ed. Pierre Girard, Christian Leduc and Mitia Rioux-Beaulne (Paris, 2016), p.171-90.

of space through reflection? Does it mean that space is an intellectual and innate idea? Also, how could space have an ontological status if we are not able to perceive it in sensation? I think that these questions led Euler to leave aside some aspects of the problem in texts such as the *Anleitung zur Naturlehre* and the *Lettres à une princesse.* It may be that Euler himself remained unconvinced by his 'Réflexions'. Perhaps this is the reason why he never returned to these arguments in his later works.

Experimental-Metaphysik at the Berlin Academy: the odd alliance between experience, mathematics and teleology

PAOLA BASSO

The three souls of the Academy's metaphysics

This chapter explores the *hybridisation* of philosophical investigation at the Berlin Academy by focusing on three of its outstanding members, each coming from different backgrounds. They are Maupertuis (1698-1759), the Newtonian and perpetual president; Euler (1707-1783), the anti-Wolffian prince of mathematicians, director of the mathematical class; and Lambert (1728-1777), the self-taught scientist-philosopher. One French and two Swiss savants under the Prussian sky.

As is well known, in restructuring the Academy, the former classes were reorganised and renamed. The first class, which covered much of natural philosophy, was called experimental philosophy;[1] the second was that of mathematics, which included also mechanics and astronomy; the third, speculative philosophy, covering metaphysics, logic and morals; and, lastly, the fourth class was that of literature.[2] Despite this division into different classes, their mutual interdependence is continually emphasised in the Academy's 'Mémoires',

The content of this chapter intersects with my 'Lambert on experience and deduction', in *The Experimental turn in eighteenth-century German philosophy*, ed. Karin de Boer and Tinca Prunea-Bretonnet (New York, 2021), p.181-202.

1. Newton used the specific term 'experimental philosophy' in the second edition of his *Principia* (1713) as opposed to 'hypothetical philosophy', in order to defend his work against Leibniz; see Alan E. Shapiro, 'Newton's experimental philosophy', *Early science and medicine* 9:3 (2004), p.185-217 (185-87). For a history of this term, see Peter R. Anstey and Alberto Vanzo, 'The origins of early modern experimental philosophy', *Intellectual history review* 22:4 (2012), p.499-518.
2. See *Histoire de l'Académie royale des sciences et belles lettres* (1746; Berlin, Haude, 1748), p.3-4.

from Formey's 'Preface' to the first volume of the *Histoire de l'Académie royale des sciences et des belles lettres* to Lambert's 'Discours de réception', pronounced twenty years later.[3]

By assembling distinguished mathematicians, natural philosophers and speculative philosophers under one roof and encouraging them to join in the collaborative *esprit académique*, the Berlin Academy housed the Newtonian pairing – an experimental inclination and a strong faith in mathematics – combined with the urge for a deep reform in metaphysics, spurred on by a growing intolerance towards sterile systems or *raisonnemens illusoires.* In this regard, the strong contamination of German speculation with English empiricism and experimentalism, traceable in the Berlin Academy, seemed to be strengthened by this extraordinary coexistence of a speculative class in the middle of a scientific society. This *espèce de singularité*,[4] to use Formey's words, will facilitate a sort of 'reshaping' of metaphysical knowledge by the Academy.

Maupertuis, Euler and Lambert, indeed, derived their ability to merge the two plans, experimental and speculative, from the twofold nature of their task. Lambert, a true metaphysician inclined to science is, perhaps, the one in which such hybridisation is more visible.[5] Nevertheless, Maupertuis and Euler, in spite of their strong anti-dogmatism, resorted to metaphysics every time natural philosophy ground to a halt, even adopting metaphysical final causes, albeit from a scientific or a mathematical point of view. The development of a 'higher science' (*science supérieure*) was still a widespread aspiration. Lambert's own *Architectonic*, for instance, is labelled by him as a 'completely new enquiry in metaphysical ontologies [*metaphysischen*

3. [Jean H. S. Formey], 'Préface', in *Histoire de l'Académie royale des sciences et des belles lettres de Berlin* (1745; Berlin, Haude, 1746), sig.)( )(2. I am grateful to Peter Anstey for flagging this work to me and for his critical reading of my chapter.
4. Formey, 'Préface', sig.)( )(1*v*: 'Besides this, we shall find in our *Mémoires* the examination of certain questions, which do not usually enter into such works. These are the questions of metaphysics, natural rights and morals.' In his view, during the scientific revolution, metaphysics, instead of offering a 'Dictionnaire de termes barbares' as in the previous scholastic age, became a kind of 'nursery [*pépiniere*]'. All translations are my own unless otherwise stated.
5. In Lambert, the *a priori* cognitions achieved by reasoning come on the trail of some undoubted *a posteriori* simple concepts and propositions achieved by experience. Accordingly, the truths that are established *a priori* do not cancel out the experimental side of scientific knowledge and remain bound to it. This hybrid *a priori* allows him to escape rigid dichotomies. On this, see Basso, 'Lambert on experience and deduction', p.181-202.

Grundlehren]',[6] and in Euler we find the quest for 'a grand science, which we yet lack [...]: the true metaphysics'. Willing or not, all parties could be for metaphysics what Copernicus was for astronomy, to use 'a parallel'[7] coined in those halls. We can also note how the hybrid status of academic metaphysics seems to mirror astronomy, the Berlin Academy science *par excellence*[8] – a mathematical science intertwined with experimental observations and cosmological assumptions.

As a means of examining the tight interconnections among the different disciplines at play in the Academy, it is appropriate to start with Lambert's 'Discours de réception' addressed to an Academy plenary meeting on 24 January 1765 (the king's birthday).[9] Here he sets out his strong paradigm of synergy (*liaison*) between the four classes, by showing their need for 'mutual help' (*secours mutuels*), considering not only 'the close connection [*l'étroite union*] between mathematics and experimental physics', but also the fact that 'physics is no less linked to philosophy'.[10] He therefore introduces his point very clearly: 'so here is philosophy, closely related to experimental physics.'[11] This is a clear acknowledgement that speculative philosophy must be shaped on the shores of experimental philosophy, or, in Merian's words, seeing metaphysics as a 'physics of immaterial beings'.[12]

6. Johann H. Lambert, *Anlage zur Architectonic* (Riga, Hartknoch,1771), p.iii.
7. '[W]ithout expecting to be a good *Metaphysician* as *Copernicus* was a good Astronomer [...] I do not fail to have in this parallel infinite advantages'; Pierre Le Guay de Prémontval, 'De la psychocratie' (part 3), in *Histoire de l'Académie royale des sciences et belles lettres* (1761; Berlin, Haude and Spener, 1768) p.371-420 (417). As rightly stressed by Schliesser: 'in the wake of the *Principia*'s success, Newton's authority is used to settle debates within philosophy and to change the character of philosophical theorizing'; see Eric Schliesser, 'Newton and Newtonianism', in *The Routledge companion to eighteenth-century philosophy*, ed. Aaron Garrett (New York, 2014), p.62-90 (76-77).
8. A primacy deriving also from the practical need, thanks to the Academy's monopoly on the making of calendars. The 'Diplome de la Fondation' of the Academy states: 'However, we will not defer too long the procuring of a skilful astronomer, who would attend to the observations of the celestial movements, and to the composition of the calendars'; *Histoire de l'Académie royale des sciences et belles lettres* (1751; Berlin, Haude and Spener, 1752), p.30.
9. Johann H. Lambert, 'Discours de réception', in *Histoire de l'Académie royale des sciences et belles lettres* (1765; Berlin, Haude and Spener, 1767), p.506-14.
10. Lambert, 'Discours de réception', p.510.
11. Lambert, 'Discours de réception', p.511.
12. 'Physique des êtres immatériels'; Merian, 'Discours sur la métaphysique', in *Histoire de l'Académie royale des sciences et belles lettres* (1765; Berlin, Haude and Spener, 1767), p.450-74 (459).

What counts is that Lambert, in an inverted perspective, begins with experimental philosophy (which he calls ***physique expérimentale***), his own class, electing it as a fulcrum for the other disciplines: 'The class of experimental philosophy [***physique expérimentale***], to which I have felt duty-bound to devote myself, will give me the opportunity to offer you a more particular point of view, because I'm going to trace back to it all other knowledge, and to show by which synergies they are connected.'[13] Praising Bacon, Lambert shows how the new challenge is to achieve metaphysical principles not as assumptions in a vacuum, but as outcomes of a scientific investigation. Therefore, natural philosophy constitutes without doubt 'the first ground of our knowledge': 'Physics [i.e. natural philosophy] expects in return the principles necessary to achieve the form of a demonstrative science.'[14]

The experimental level not only grants an *a posteriori* beginning, but also provides a sort of *fil rouge* to follow.[15] But this is not enough: the second class is also required, mathematics (i.e. the science of quantity and demonstration). Only by examining the dependence and subordination of ideas can we forge the right principles or laws, reaching 'the reasoning that follows geometrical order and rigour'.[16] In an almost Kantian formula, Lambert points immediately to the very peculiarity of the metaphysical enquiries at the Berlin Academy, the combination of experience and calculation: 'By neglecting calculation and the theory on which it is based, experiments are made without choice or design. By neglecting experiments, one runs the risk of falling into a chimera, and of producing calculations applicable to any other

13. 'La Classe de Physique expérimentale, à laquelle j'ai cru devoir me vouer, me donnera lieu de Vous offrir un point de vue plus particulier, en ce que je vais ramener les autres connoissances à celle-ci, & faire voir par quelles liaisons elles y tiennent'; Lambert, 'Discours de réception', p.506.
14. 'La Physique donne à la Philosophie les matériaux, toutes les premieres idées & connoissances. Mais elle en attend en échange les principes nécessaires [...] pour prendre la forme de Science démonstrative'; Lambert, 'Discours de réception', p.511.
15. In this regard, see Lambert's claim of 'a supply of concepts and propositions' expressly drawn from *experience* (i.e. simple concepts and axioms) in order to subsequently form *complex* concepts and propositions that can apply to the real world, although not directly based on experience; see Johann H. Lambert, *Neues Organon oder Gedanken über die Erforschung und Bezeichnung des Wahren und dessen Unterscheidung vom Irrtum und Schein* (Leipzig, Johann Wendler, 1764), reprinted in *Philosophische Schriften*, vol.1 (Hildesheim, 1965), §577, p.367.
16. Lambert, 'Discours de réception', p.510.

world, but the one in which we are.'[17] To this second class, Bacon's or Locke's achievements no longer seem enough: 'Mr. Locke began by disentangling the simple ideas, but without applying the method I have just indicated' (i.e. 'the reasoning which follows geometrical order and rigour').[18] The model of reasoning offered by mathematics remains opposite to that offered by experience, but – after Newton and Galileo – it is acknowledged that they can lead to the same results.

In order to turn this alliance into a unified system, we need the third class, the specific and peculiar class of this Academy, that is, speculative philosophy, which, building on the first two, experimental physics and mathematics, endows them with explanatory connections and general principles: 'These universal principles, generally applicable, with which metaphysics ought to furnish us, and from which we derive a great help, when it is a question of *linking* experiments and of giving physics *a truly scientific order and connection*.'[19] This mutual interdependence is clearly bidirectional: 'experimental physics' and mathematics make possible *la véritable métaphysique*, while metaphysics, with its foundational tenets, makes experimental physics *véritablement scientifique*. It is worth noting, at this juncture, that Euler had already pointed out the regulative role of natural sciences for metaphysics, in his 'Réflexions sur l'espace et le tems': 'it will be on these [i.e. mechanical] *conclusions* that one will have to adjust and establish the first ideas of metaphysics', in fact 'it is therefore always a great advance, when we already know some conclusions, to which the first principles of metaphysics must lead.'[20]

17. 'En négligeant le calcul, & la théorie qui lui sert de base, on fait les expériences sans choix & sans dessein. En négligeant les expériences, on court le risque de tomber dans la chimère, & de produire des calculs applicables à tout autre Monde qu'à celui où nous sommes'; Lambert, 'Discours de réception', p.509.
18. 'Mr. Locke commença par débrouiller les idées simples, mais sans y appliquer la méthode que je viens d'indiquer'; Lambert, 'Discours de réception', p.511.
19. '[C]es' principes universels, généralement appliquables, que nous doit fournir la Métaphysique, & dont on tire un grand secours, lorsqu'il s'agit de *lier* les expériences & de donner à la Physique *un ordre & une connexion véritablement scientifique*'; Lambert, 'Discours de réception', p.511 (my emphasis).
20. 'C'est donc toujours une grande avance, quand on connoit déjà d'ailleurs quelques conclusions, auxquelles les premiers principes de la Metaphysique doivent aboutir: & ce sera sur ces conclusions, qu'il faudra régler & déterminer les premieres idées de la Metaphysique'; Leonhard Euler, 'Réflexions sur l'espace et le tems', in *Histoire de l'Académie royale des sciences et belles lettres* (1748; Berlin, Haude and Spener, 1750), p.324-33 (325). Euler, despite being aware of the bad influences of Cartesian metaphysics in natural philosophy, was also

Euler, introducing mechanics '*as a guide* in this thorny research',[21] is perhaps more extreme than Lambert in giving supremacy to natural science over and above metaphysics, but both aim to free metaphysics from its isolation.

With this project of introducing experimental evidence into the core of metaphysics, we are facing more than a 'priority problem'[22] between physics and metaphysics: it is a reshaping of metaphysics itself, in this age of deep osmosis between science and philosophy. This deep synergy between the first three classes of the Academy introduced a new scenario where the disciplines, while keeping their respective methods and degrees of certainty well separated, are called upon to pool their *instruments* and *results*: 'they all begin from the same starting point. The difference consists only in the path, and, if I may use the term, in the direction of the different roads.'[23]

And yet, in the end, these 'different roads' are not supposed to contradict each other: as Maupertuis put it in 1743 at the beginning of his 'Accord de différentes loix de la nature': 'it would be overwhelming to find out that the propositions that philosophy gives us as fundamental truths, were disproved by geometrical reasoning or algebraic calculations.'[24] This statement would be strengthened some years later by Euler, who authorised the 'rejection' of all the 'reasoning and ideas' in metaphysics 'that lead to conclusions contrary to these truths [of mechanics]'.[25] More than an indictment of metaphysics,

---

aware of their synergy: 'Also metaphysicians, far from denying these principles, the truth of which mechanics assures us, rather endeavour to deduce and demonstrate them by their ideas'; Euler, 'Réflexions sur l'espace', p.325.

21. 'La connoissance de ces verités pourra servir de guide dans ces recherches épineuses'; Euler, 'Réflexions sur l'espace', p.324.
22. On the priority problem, see Christian Leduc, 'La métaphysique de la nature à l'Académie de Berlin', *Philosophiques* 42:1 (2015), p.11-30 (30), who argues that, in Euler and Maupertuis, metaphysics is no longer understood as foundational with respect to the natural sciences.
23. 'Elles partent toutes des mêmes commencemens. La différence ne consiste que dans la marche, & si j'ose employer ce terme, dans la direction des différentes routes'; Lambert, 'Discours de réception', p.510.
24. 'Il seroit accablant de voir que des propositions que la Philosophie nous donne comme des vérités fondamentales, se trouvassent démenties par les raisonnemens de la Géométrie, ou par les calculs de l'Algèbre'; Pierre-Louis M. de Maupertuis, 'Accord de différentes loix de la nature', in *Œuvres de Maupertuis*, 4 vols (Lyon, Bruyset, 1768), vol.4, p.1-28 (3-4).
25. 'On sera en droit de rejetter tous les raisonnemens et toutes les idées [...] qui conduisent à des conclusions contraires à ces verités'; 'le Metaphysicien sera obligé

these demands are meant to achieve a new metaphysics, compatible with the results of mathematics and natural philosophy.[26]

It was, therefore, still deemed possible to reconcile rational metaphysics and experimental physics. Such continuity is granted if all abstract concepts are rooted in this world.[27]

Wolff himself had already tried to pursue such a project, appealing to both mathematics and experience,[28] but he seemed to have missed the mark, searching for the uniformity of a universal method and completely subordinating experience to metaphysics.[29] In this new scenario, by contrast, physics and mathematics are called upon to verify metaphysics, no longer in a subordinate position but as *pares inter pares*,[30] in a sort of handover: the motto indeed was 'metaphysics

---

de remédier à ce défaut, & de substituer dans l'énonciation de ces principes des idées réelles au lieu des imaginaires'; Euler, 'Réflexions sur l'espace', p.324-25.

26. It has been said of Maupertuis that he was 'halfway between the mermaid of metaphysics and positivist prudence'; see Jean Erhard, *L'Idée de nature en France dans la première moitié du XVIIIe siècle* (Paris, 1963), p.172. In his 'Discours prononcé le jour de la naissance du roi (1747)', Maupertuis, to address the members of this Academy, uses the following circumlocution: 'Physicien, Géomètre, Philosophe, Orateur', in *Histoire de l'Académie royale des sciences et belles lettres* (1746; Berlin, Haude, 1748), p.10-16 (16).

27. For Lambert the *intellectual world* (*Intellektualwelt*) should be built upon a *material world* (*Körperwelt*): in order to get to the intellectual world, it is important to follow the firm track of the material world as a regulative 'metaphor'; if one reaches the intellectual world too quickly, 'one often ends up in a flight of Icarus [*un vol d'Icare*] that results in a fatal fall'; Johann H. Lambert, 'Observations sur quelques dimensions du monde intellectuel', in *Histoire de l'Académie royale des sciences et belles lettres* (1763; Berlin, Haude and Spener, 1770), p.421-43 (431). Lambert grounds his ontology on the concept of *solidity* and *force* (see the 'Alethiology' section of his *New Organon* and the third part of his *Architectonic*: 'The real of ontology'); as Beck points out, these two concepts made it possible to cross 'the chasm between logical and metaphysical truth'; see Lewis W. Beck, *Early German philosophy: Kant and his predecessors* (Cambridge, 1969), p.203.

28. Wolff counts 'undoubted [*ungezweifelte*] experiences' among the first principles of an axiomatic system, thus putting them on a par with definitions and axioms; see, for instance, Christian Wolff, *Vernünfftige Gedancken von der Kräfften des menschlichen Verstandes* (Halle, Renger, 1713), ch.4, §21. On this see Corey W. Dyck, *Kant and rational psychology* (Oxford, 2014), p.40.

29. 'Maupertuisian cosmology is in fact an *a posteriori* theology, based on a teleological interpretation of the existing world order. The opposition to Wolff could not be sharper'; Giorgio Tonelli, *La Pensée philosophique de Maupertuis* (Hildesheim, 1987), p.24.

30. While Béguelin, Formey and Lambert symbolise 'the happy alliance of the *esprit philosophique* with the *esprit mathématique*' (Christian Bartholmèss, *Histoire philosophique de l'Académie de Prusse depuis Leibniz jusqu'à Schelling*, 2 vols, Paris,

starts where physics stops.'[31] A case in point of a '*metaphysical* problem' – defined by D'Alembert as 'le grand Probléme *Métaphysique*'[32] – was the question proposed by Euler on behalf of the class of experimental philosophy (for the year 1758).[33]

But what sort of metaphysics did such an intellectual milieu give rise to?

A name for this hybrid: 'experimental metaphysics'

It is time to focus on a term referring precisely to Lambert's and Euler's way of philosophising, in contrast to Wolff's way. The term was devised by Georg Jonathan von Holland, mathematician and philosopher and corresponding member of the Berlin Academy from 1749. Describing the peculiar philosophical method of his interlocutor in a letter to Lambert of 22 September 1765, Holland expressly uses the rich neologism of 'experimental metaphysics': 'We gain therefore an *experimental metaphysics*: a science nobody has contemplated yet.'[34]

The term 'experimental metaphysics'[35] emerges in an anti-Wolffian context, and Newton is cited as a model:[36]

1850, vol.2, p.2), Maupertuis, Euler and Merian condemned that alliance when intended as a claim to a common methodology: 'this affectation of the geometrical method has done everlasting harm to the speculative sciences' (Merian, 'Discours sur la métaphysique', p.460), but not as a mere collaboration.

31. 'La Métaphysique commence là où la Physique finit'; Merian, 'Discours sur la métaphysique', p.459.
32. Jean D'Alembert, *Traité de dynamique*, 2nd edn (Paris, chez David l'aîné, 1758), 'Discours préliminaire', p.xxiv.
33. 'Si les loix de la Statique et de la Méchanique sont de verité necessaire ou contingente?': according to D'Alembert, to answer this question, one has to enquire 'si les loix données par l'expérience s'accordent avec celles que le raisonnement seul a fait trouver', or 'si les unes et les autres sont différentes' (*Traité de dynamique*, p.xxvi-xxvii). On this question see André Charrak, *Contingence et nécessité des lois de la nature au XVIII*[e] *siècle: la philosophie seconde des Lumières* (Paris, 2006).
34. 'Wir erlangen dadurch eine Experimental-Metaphysik; eine Wissenschaft, an die man bisher nicht gedacht hat'; Holland to Lambert (22 September 1765), Letter IX, in Johann H. Lambert, *Briefwechsel*, in *Philosophische Schriften*, vol.9 (Hildesheim, 1968), p.92.
35. Although the term 'metaphysics' could have a very general meaning such as 'philosophy', the fact that Holland speaks of a science that 'nobody has contemplated yet' shows how he is willing to deal with something different from the already customary term, i.e. 'experimental philosophy'. The term *Experimental-Metaphysik*, written with the hyphen, seems to mimic the Wolffian term *Experimental-Physik*.
36. Although we are dealing here with an advocated dissociation from the Wolffian

> One proceeds in it [in experimental metaphysics] as in natural philosophy [*Naturlehre*]. One collects simple phenomena by means of observations, and calculates their relationships. Meanwhile, one can leave aside the causes of simple phenomena, as *Newton* did, for instance, with the cause of attraction; and even if men should be so lucky as to discover them, the propositions found by the experimental metaphysics [*die durch Experimental-Metaphysik gefundene Sätze*] must suffer no change.[37]

The novelty, here, is to require metaphysics to align itself with natural philosophy. From the 'eternal geometrician' metaphysician to the 'natural philosopher' metaphysician, the step is not short, but inevitable: to avoid the many chimeras with which metaphysics was becoming synonymous. We read in the 'Mémoires' of the Academy: 'the perfect metaphysician [...] would be in turn the perfect natural philosopher'[38] – all the members of the Academy seem to be involved in this reform.

What would we be left with, if metaphysics were not aligned with physics? Holland puts it clearly: 'If we had left the *theory of motion* in the hands of metaphysicians without pushing it forward [...] we would not have gone beyond Zeno!'[39] This is the reason why experimental science and metaphysics must become allies: here is the ground for the construction of an 'experimental metaphysics' in opposition to a purely abstract metaphysics, whose archetype was indeed Zeno, because he totally banished experience from his speculations. As Lambert once wrote to Kant, in 1770: 'in my opinion, a very genuine metaphysician [*ein purus putus metaphysicus*] is like someone lacking a sense, like a blind man lacking sight.'[40] Compared with the new

method, we have to stress the circumstance that Holland and Lambert were both involved with a Leibnizian *forma mentis* and a Wolffian education.

37. Holland to Lambert (22 September 1765), in Lambert, *Briefwechsel*, p.92.
38. 'Non-seulement le parfait Métaphysicien posséderoit des connoissances plus utiles, plus profondes, que ne le seroit la Physique; il seroit parfait Physicien à son tour'; Merian, 'Discours sur la métaphysique', p.455.
39. Holland to Lambert (22 September 1765), in Lambert, *Briefwechsel*, p.92. An assumption that recalls what was stated in 1750 by Euler: 'Les Metaphysiciens ont donc tort, quand ils veulent bannir entierement du monde l'espace & le lieu, en soutenant que ce ne sont que des idées abstraites & imaginaires' (Euler, 'Réflexions sur l'espace', p.329).
40. Lambert to Kant (13 October 1770), in Lambert, *Briefwechsel*, p.356. *Purus putus* (or *purus ac putus*) is the technical expression for the purity of gold and silver and means 'unalloyed'.

experimental metaphysician, Zeno acted like a blind man! The era of the dogmatists spinning webs out of themselves was definitively over. Hypotheses and pure thought seem to have fallen into disgrace: even Leibniz, driven by the spread of scientific instruments, once wrote to Huygens: 'I prefer a Leeuwenhoek who tells me what he sees, to a Cartesian who tells me what he thinks.'[41]

The term 'experimental metaphysics' could sound like an oxymoron, a paradoxical expression, in the light of the fact that metaphysics in itself used to hint at something intrinsically speculative, which goes beyond the domain of experience, and also for the intense opposition that experimental philosophers had to metaphysics, but it is not. Holland wields this new science against the demand, attributed to Wolff, of deducing whole systems from empty definitions: 'One will see that Wolff had abstracted away from the simple things to such an extent that basically nothing remains.'[42]

As Holland further explains, the methodological mistake lies in the metaphysical claim of knowing a thing in its *totality*, while the mathematician or the physicist does not consider the whole of a thing, but rather its single aspects separately: the shape, the weight, the size, the force and so on. Therefore, the new experimental metaphysician should approach things in this scientific way, moving step by step from single observations. Proceeding like Lambert in his 'Discours de réception', after having included natural philosophy, Holland adds another ingredient, namely, mathematics, quoting the enterprises of Euler: 'The mathematician has put aside all the unnecessary questions and, by observations and calculation [*durch Beobachtung und Rechnung*], has produced a science extremely useful for mankind. Euler brings this work nearly to perfection; whereas the *metaphysicians until now* still fight over whether something like body, movement, time, space, etc., exists or not.'[43] Here we come back to the pair *esprit d'observation* and *calcul*, a union never undertaken by metaphysicians;

41. 'Je préfère un Leeuwenhoek qui me dit ce qu'il voit, à un Descartes qui me dit ce qu'il pense'; Leibniz to Huygens (2 March 1691), in Christiaan Huygens, *Œuvres complètes de Christiaan Huygens*, 22 vols (The Hague, 1888-1950), vol.10, p.52. It became a commonplace to claim: 'on s'en assûre par les yeux.'
42. Holland to Lambert (22 September 1765), in Lambert, *Briefwechsel*, p.80, against abstract metaphysical systems based on the empty concept of *ens*. See, in this regard, Wolff's idea that ontology should 'deal with the most abstract notions': Christian Wolff, *Philosophia prima, sive Ontologia* (Frankfurt and Leipzig, Renger, 1729), §26.
43. Holland to Lambert (22 September 1765), in Lambert, *Briefwechsel*, p.92-93.

but here the blame seems to be put on the metaphysicians[44] as supporters of sterile controversies, and not necessarily on metaphysics itself. Instead, Lambert's 'manner of philosophising' (i.e. experimental metaphysics) 'is far more appropriate to our nature than any other method'.[45] Concerning this ***experimental metaphysics***, it is really striking how this 1765 Holland letter to Lambert is similar, in language and purpose, to a letter from Euler to Maupertuis from 1745, in which Euler explicitly invites metaphysics to incorporate the principles of physics and mathematics for the birth of a completely new science:

> Here is, then, *a grand science*, which we yet lack, and which deals with the general principles which are observed in nature, and it seems to me that that is the very place where the *genuine metaphysics* resides, as far as it contains the first principles of *physics* and *mathematics*; from which the metaphysics of Leibniz and Wolff is still very far removed.[46]

We find the same contrast with Wolff and almost the same words: the expression 'the genuine metaphysics [...] which we yet lack' evokes Holland's '*experimental metaphysics*: a science nobody has contemplated yet'. Lastly, the third of our authors, Maupertuis, entitles the seventeenth paragraph of his *Lettre sur le progrès des sciences* 'Expériences métaphysiques' (i.e. 'Metaphysical experiments'). Here the term 'metaphysics' implies that it is a mere counterpart of physics: if physics deals with bodies, metaphysics deals with minds or immaterial beings. In fact, D'Alembert himself, 'the secret president', in his *Eléments de philosophie*, had assigned *matière* to physics and

44. Euler, for instance, used to deride the Wolffian philosophers as *les métaphysiciens*. The same disposition can be found in Maupertuis, who, while complaining to Johann II Bernoulli that 'La métaphysique d'Allemagne est une étrange science', felt compelled to add: 'mais ce n'est pas la faute de la métaphysique, c'est la faute des Allemands'; Maupertuis to Jean II Bernoulli (18 September 1747), Bibl. Bale, L.Ia 708, n 102, quoted in *Commercium epistolicum*, vol.6: *Correspondance de Leonard Euler avec de Maupertuis et Frédéric II*, in *Opera omnia*, ed. Pierre Costabel, Eduard Winter, Ašot T. Grigorijan and Adolf P. Juškevič, series 4.a (Basel, 1986), 'Introduction de P. Costabel', p.6.
45. Holland to Lambert (22 September 1765), in Lambert, *Briefwechsel*, p.92-93.
46. 'En voilà donc *une grande science*, qui nous manque encore, et qui roule sur les principes generaux qui s'observent dans la nature, et il me semble que c'est là où réside la *veritable metaphysique*, en tant qu'elle renferme les premiers principes de la *physique* et de la *mathematique*; de laquelle la metaphysique de Leibniz et de Wolff est encore bien éloignée'; Euler to Maupertuis (10 December 1745), in *Commercium epistolicum*, p.57 (my emphasis).

esprit to metaphysics.[47] What is striking here is simply the idea that something like 'metaphysical experiments' could exist.[48]

But to what extent does experimental metaphysics remain a metaphysics and not transform itself into science? This brings us to the third, strictly metaphysical, ingredient of the Academy's metaphysics, which can be regarded as the integration of a preliminary scientific procedure, and which should come to the fore as soon as the other two tenets, experience and mathematics, fall down – that is, its intrinsic teleological inclination.[49] Euler himself will make a strong use of this addition.

The best example of the hybridisation at play in the Berlin Academy is indeed the collegial reworking of Maupertuis's Principle of least action, a law, more than an axiom, derived from experience, confirmed by mathematics, but necessarily embedded in a metaphysical framework. As is well known, Maupertuis invokes this Principle of least action, according to which, when a change happens in nature, the necessary action required is the smallest one. This principle, notwithstanding the limitations of its original disciplinary domain (dioptrics), was shown to be universal by Euler, who applied it successfully to dynamics (i.e. the study of the trajectories of bodies): 'I also believe your principle to be more general than you propose.'[50] In light of this success, we can witness Euler's endorsement of Maupertuis's teleology directly in his mathematical masterpiece, his *Methodus inveniendi*: 'Therefore, a twofold path [*duplex via*] to understand the effects of Nature lies open to us: one by efficient causes, which is usually called the direct method; the other by final causes; the

47. Facing the 'question of what is left remaining for Metaphysics', Bacon answered that 'Physicke should contemplate that which is inherent in Matter, and therefore transitorie; and Metaphysicke, that which is abstracted and fixed'; Francis Bacon, *The Advancement of learning*, ed. M. Kiernan, in *The Oxford Francis Bacon*, vol.4 (Oxford, 2000), p.82.
48. In this case, Maupertuis deals with experiments about mind, perceptions and about the origin of language (a subject deeply felt by the Berlin Academy, which proposed many prize questions thereupon).
49. A metaphysics which is teleological, notwithstanding the strong opposition to *particular* arguments from design.
50. Euler to Maupertuis (10 December 1745), in *Commercium epistolicum*, p.56. On Euler and Maupertuis, see also Marco Panza, 'De la nature épargnante aux forces généreuses: le principe de moindre action entre mathématiques et métaphysique – Maupertuis et Euler, 1740-1751', *Revue d'histoire des sciences* 48:4 (1995), p.435-520.

mathematician uses both with equal success.'[51] We can find an antecedent of this bizarre modern double-truth in Bacon himself, when he writes that final causes, abolished in physics (and condemned as 'sterile investigation'), are nonetheless admitted in metaphysics.[52]

We know all the warnings Maupertuis places before 'the need to posit an *intelligent cause* in the teleological determination of things', opening the door again to *final causes*: 'it is not without risk that we introduce them.'[53] With fewer caveats, Lambert, quoting Maupertuis's teleology explicitly in his *Cosmological letters* (*Cosmologische Briefe über die Einrichtung des Weltbaues*), describes his purposefully arranged cosmos: 'Mr. von Maupertuis also endeavoured to demonstrate all laws of motion from teleology.'[54] He went so far as to state: 'it is regrettable that the study of teleology is still a matter of sheer faith [*blosse Glaubensache*].'[55] However, this new teleological attitude could resurface only under the guarantee of its 'mathematisation'.

Leibniz had already introduced the idea of a teleological principle of *simplicity*, but only the triad Maupertuis–Euler–Lambert, perhaps inspired by cosmology, dared to build such a deep link between science and teleology. Indeed, Lambert, in his *Cosmological letters*, more than in other writings, is unequivocally teleological, and confesses to having written that book 'to find in the whole world-edifice the kind of order,

51. Leonhard Euler, *Methodus inveniendi lineas curvas maximi minimive proprietate gaudentes, additamentum* 1: *De curvis elasticis* (Lausanne and Geneva, Bousquet, 1744), p.245-310 (245), reprinted in *Opera omnia*, ed. A. Speiser, series 1, vol.24 (Bern, 1952).
52. 'For as wee devided *naturall philosophy* in *generall* into the *enquirie* of *causes* & *productions* of *effects*: So that part which concerneth the *enquirie* of *causes*, wee do subdivide, according to the received and sound division of *causes*; The one part which is *phisicke* enquireth and handleth the *materiall* & *efficient causes*, & the other, which is *metaphysicke* handleth the *formal* and *final causes*' (Bacon, *The Advancement*, p.82). Here metaphysics is understood as a mere part of natural philosophy, whose enquiries are not mechanical but teleological.
53. 'J'avoue que ce n'est pas sans peril qu'on les introduit'; Maupertuis, 'Accord de différentes loix', p.20. Elsewhere Maupertuis wrote: 'It is therefore not in the small details that we must seek the Supreme Being: it is only in the phenomena the universality of which does not suffer any exception and the simplicity of which is entirely exposed to our view'; Pierre-Louis M. de Maupertuis, 'Les loix du mouvement et du repos déduites d'un principe métaphysique', in *Histoire de l'Académie royale des sciences et belles lettres* (1746; Berlin, Haude, 1748), p.267-94 (277).
54. Johann H. Lambert, *Cosmological letters on the arrangement of the world-edifice*, translated by Stanley L. Jaki (New York, 1976), preface, p.46.
55. Lambert, *Cosmological letters*, p.46.

harmony, variety, transformation, interconnection, perfection, beauty, means, and purpose which we admire on the earth'.[56] Well accepted inside the Academy, these letters were translated and summarised in an extract by Merian, under the title of *Système du monde* (1770). As Maupertuis himself acknowledged in the preface of his own *Essai de cosmologie*: 'if even a Newton has left here [in cosmology] so much to be desired [with regard to teleology], who will be the man who dares to undertake it?'[57] After all, in Germany, despite the publication of the Leibniz–Clarke correspondence,[58] Newton was never seen as a metaphysician and, in fact, 'Newtonian metaphysics was the brainchild of Newton's disciples.'[59]

Holland's challenge addressed to Lambert is therefore highly redolent: the results of experience, calculation and God's design converge in a peculiar 'experimental metaphysics' which, like every hybrid, implies some odd alliances.

The odd alliance between experience and mathematics

In the coupling of experience and calculation, that is, in the mathematisation of natural philosophy,[60] philosophy gets verification and

56. Lambert, *Cosmological letters*, p.74.
57. 'Si un Newton y a laissé tant de choses à desirer, quel sera l'homme qui osera l'entreprendre?'; Pierre-Louis M. de Maupertuis, *Essai de cosmologie*, in *Œuvres de Maupertuis*, vol.1, p.1-78.
58. The Leibniz–Clarke correspondence (Samuel Clarke, *A Collection of papers, which passed between the late learned Mr. Leibniz, and Dr. Clarke, in the years 1715 and 1716*, London, James Knapton, 1717) was published in a German translation in 1720, prefaced by Wolff and edited by Thümming.
59. Thomas Ahnert, 'Newtonianism in early Enlightenment Germany, c. 1720 to 1750: metaphysics and the critique of dogmatic philosophy', *Studies in history and philosophy of science* 35:3 (2004), p.471-91 (473). As Ludwig Kahle wrote in 1741 against Voltaire, despite his book's title *A Comparison of the Leibnizian and Newtonian metaphysics*: 'Newton remains a great man, even if he was no metaphysician'; see Ludwig M. Kahle, *Vergleichung der Leibnitzischen und Neutonischen Metaphysik* (Göttingen, Königl. Universitäts-Buchhandlung, 1740), preface, unpaginated.
60. Not only were the canonical sciences to be mathematised, but also the rest of human knowledge. See in this regard, Wolff's idea of a 'Psychometrie' and Maupertuis's mathematical estimation of moral sentiments: 'l'estimation des momens heureux ou malheureux est le produit de l'intensité du plaisir ou de la peine par la durée' and 'Le bonheur est la somme des biens qui reste après qu'on a retranché la somme des maux'; Pierre-Louis M. de Maupertuis, *Essai de philosophie morale*, in *Œuvres de Maupertuis*, vol.1, p.171-252 (195 and 184-85).

heuristic power from the former (as certified by Lambert's 'Discours de réception': 'experience *verifies* both, and *discovers* every circumstance omitted, or falsely admitted')[61] and universality, necessity and rigour from the latter. Experience and calculation become twin instruments for the human mind, that is to say, epistemological tools. After Newton, this pairing was incontestable, to the point that it came to be out of the question to accept as philosophical truth something which contradicts observations or mathematical truths. It is, in fact, not a coincidence that Lambert devoted some chapters of his *Architectonic* to 'lovers of natural philosophy' (*Liebhaber der Naturlehre*) and others to 'lovers of mathematics' (*Liebhaber der Mathematik*): a commitment which, by his admission, 'can scarcely be seen in other metaphysics'.[62] Holland, presenting his idea of 'experimental metaphysics', calls the natural philosopher a mathematician as well. Euler also speaks often of the 'marvellous accord' between mathematics and physics,[63] and Maupertuis himself, looking for someone 'capable of supporting the metaphysical class', searches explicitly for 'someone broad-minded, having *mathematical* knowledge'.[64]

Furthermore, it is important to stress that *experience* in the Academy is regarded no longer as the mere 'occasion' of knowledge – its blind origin – but as an essential element, thanks also to the homonymy with the term 'experiment' in French. Through its power of consulting

61. 'L'expérience vérifie l'une et l'autre, et découvre chaque circonstance omise, ou faussement admise' (Lambert, 'Discours de réception', p.509). 'Deux exigences sont formulées pour constituer le *véritable physicien*, du côté expérimental comme du côté mathématique, ce qui, évidemment, combine les intérêts scientifiques des deux premières classes'; Frédéric de Buzon, 'Quatre essais académiques: Johann Heinrich Lambert', *Les Cahiers philosophiques de Strasbourg* 44 (2018), p.127-48 (130).
62. Lambert, *Anlage zur Architectonic*, preface, p.x.
63. Euler acknowledged: 'tout est fondé sur des *expériences* les mieux constatées, et sur des principes dictés par la *raison*'; see Leonhard Euler, *Lettres à une princesse d'Allemagne sur divers sujets de physique et de philosophie*, vol.2 (St Petersburg, Académie impériale des sciences, 1768), vol.2, Letter 80, p.1. Difficulties arise when those two safe tenets are out of reach, as in the debate around the nature of space and time.
64. Bibl. Bale, L.5a 708, n 102-103 (in Euler, *Commercium epistolicum*, p.6). The linking of philosophy and mathematics was a common pattern for academies at that time (see, for instance, the titles of the periodicals of the Turin Academy of science, initially called *Miscellanea philosophico-mathematica* and thereafter *Mélange de philosophie et de mathématique de la Société royale de Turin*, edited by Lagrange, with contributions by Euler).

nature, and therefore of confirming or disproving assumptions, experience is destined to become a *final test* for philosophy. The demand of experience could follow different degrees; in Lambert's case, for instance, philosophers should go back to experience 'whenever the means of advancing by calculation and reasoning require new data',[65] whereas, for Maupertuis, 'reasoning is continually in need of experience.'[66]

But a new trend was set such that experimental tenets were to be set within a speculative frame, that is, mathematics and the force of its demonstrations; an extraordinary couple – the exactness of mathematics and the experimental side of nature – not as obvious as we consider it now,[67] foreseen by Galileo's statement that the book of nature had been written with mathematical characters. Lambert, for instance, shows his inclination to overcome dichotomies by means of his daring analogies, such as the one between the natural philosopher's procedure and the geometrician's: 'a well-devised experiment [...] can always go hand in hand with the most ingenious and difficult problem a skilled geometrician can solve.'[68]

Wishing for the odd coupling of mathematics with metaphysics, they envisage not the old Cartesian claim concerning a general geometrical method, but the regulative inclusion of mathematical results in metaphysics. Euler went so far as to want to check 'if it is always possible to substitute a metaphysical expression with a mathematical

65. 'Nous y [à la physique expérimentale] revenons, toutes les fois que le moyen de les avancer par le calcul, et par le raisonnement, requiert de nouvelles données'; Lambert, 'Discours de réception', p.513.
66. Maupertuis, *Eloge de Montesquieu*, in *Œuvres de Maupertuis*, vol.3, p.259-433 (399).
67. On the complexity of this idea of mathematisation, see *The Language of nature: reassessing the mathematization of natural philosophy in the seventeenth century*, ed. Geoffrey Gorham, Benjamin Hill, Edward Slowik and C. Kenneth Waters (Minneapolis, MN, 2016). On the shared ideology of the Royal Society intended to prevent the compelling force of mathematical demonstrations to enter the realm of natural philosophy and to restate the opposition of two philosophical camps, see Ursula Goldenbaum, 'It is not experience that matters in the battle of rationalism and empiricism, or how theology favors empiricism over rationalism', in *Leibniz's experimental philosophy*, ed. Arnaud Pelletier, *Studia Leibnitiana* 46 (2016), p.41-68.
68. 'Une expérience bien imaginée [...] peut toujours aller de pair avec le probleme le plus ingénieux & le plus difficile qu'un habile Géometre puisse résoudre'; Lambert, 'Discours de réception', p.509.

one, without falling into error'.[69] Thus, for example, he asserts: 'I imagine that in considering these cases, and the various ways in which this same formula $\int Vdv + \int V'dv' + \int V''dv''$ enters into each of them, it will no longer be so difficult to discover the principles of metaphysics which must lead to these formulas.'[70] Metaphysical principles have to lead to these mathematical formulas! It is hardly surprising then that the later Castillon's 'Essai d'une théorie métaphysico-mathématique de l'expérience' (1796) confirmed the habit of the Berlin Academy 'de soumettre au calcul des objets *métaphysiques*'[71] within an experimental framework.

It is noteworthy that in the 'flagship volume' of the *Histoire de l'Académie royale* of the completely renewed Academy (the 1746 issue), Maupertuis explicitly opts to contribute to the section of the class of speculative philosophy with his standard text on the Principle of least action.[72] The text is settled as usual, with all the mathematical formulas and experimental data, but with a focus (at least in the title) on the *metaphysical* nature of this principle.

This unforeseen alliance between mathematics and philosophy conferred new life on mathematics as well. Maupertuis hints of 'a more fitting employment of this science [*un usage plus heureux de cette science*]',[73] no longer restricted to trivial line-measurements or to hopeless endeavours like squaring the circle: 'hitherto, mathematics has had as its object only the gross needs of the body, or the useless speculations of the mind', whereas now mathematics can prove God's existence: 'the proofs of the existence of God which it will furnish will have over all the others the advantage of the evidence

69. 'Si l'on peut toujours, sans tomber en erreur, substituer l'expression metaphysique au lieu de la mathematique, de la verité de laquelle nous sommes déjà convaincus'; Euler, 'Réflexions sur l'espace', p.327.
70. Euler to Maupertuis (9 May 1748), in Euler, *Commercium epistolicum*, p.108.
71. 'On a donné le nom de *physico-mathématiques* à des traités où l'on applique les mathématiques à la physique, de même on peut désormais donner le nom de *métaphysico-mathématiques* aux tentatives qui ont pour but de soumettre au calcul des objets métaphysiques'; Jean de Castillon, 'Essai d'une théorie métaphysico-mathématique de l'expérience', in *Mémoires de l'Académie royale des sciences et belles lettres* (1796; Berlin, Decker, 1797), p.364-90 (370).
72. Mary Terrall, *The Man who flattened the earth: Maupertuis and the sciences in the Enlightenment* (Chicago, IL, 2002), p.244. Although Maupertuis refuses to label his proof of God's existence as 'metaphysical' (*Essai de cosmologie*, p.ix-x). We refer here to Maupertuis's *Mémoire*: 'Les loix du mouvement et du repos déduites d'un principe métaphysique', published in 1746.
73. Maupertuis, *Essai de cosmologie*, p.22.

which characterises mathematical truths.'[74] It becomes a question of applying mathematics to physical phenomena, which till now were known to us only empirically.

Not only the results of natural philosophy, but also mathematical results are said to play a regulative role for metaphysics. What I want to emphasise is that both experience and calculation play a crucial role in avoiding chimeras. Maupertuis, in his tireless defence of the natural philosopher, even incorporates the 'sentiment of truth', that is, the aversion to chimeras, in the statutes of the Academy, namely in his discourse 'On the duties of the academician'.[75] Philosophy, in line with D'Alembert's views, seemed to be at a crossroads: 'Philosophy is the science either of facts or of chimeras.'[76] In his 'Discours de réception', Lambert also stresses the importance of talking about the real world rather than about imagined ones in order to avoid 'giving into chimeras'.

But why does metaphysics require both observations and calculations, experience and mathematics? Precisely because metaphysics has within itself no definitive forms of verification. Euler's view is that we reach metaphysical principles only if they are proved by mathematics or experience or both (e.g. by mechanics); otherwise, it would be absurd to think that a science under suspicion of positing chimeras could subsist by itself.

It is noteworthy that, lacking direct links to proofs, metaphysics was bound to borrow a technical term from chemistry, in order to hint at a sort of external test to avoid chimeras. I am referring here to the term *pierre de touche* or, in German, *Probierstein* (in Latin, *lapis lydius*). This term was commonly used in philosophy during the eighteenth century and particularly at the Berlin Academy.[77] The

74. 'Les preuves de l'existence de Dieu qu'elle fournira auront sur toutes les autres l'avantage de l'évidence, qui caractérise les vérités mathématiques'; Maupertuis, *Essai de cosmologie*, p.22.

75. 'Cet esprit académique, cet espece de sentiment du vray, qui le lui fait découvrir partout où il est, & l'empeche de le chercher là où il n'est pas. Combien de *chimeres* qu'ils n'auroient osé produire dans une Academie!'; Pierre Louis M. de Maupertuis, 'Des devoirs de l'académicien', in *Histoire de l'Académie royale des sciences et belles lettres, depuis son origine jusqu'à présent, avec les pieces originales* (Berlin, Haude and Spener, 1750), p.115-25 (116); see the appendix to this volume.

76. 'La Philosophie [...] est la Science des faits, ou celle des chimeres'; Jean D'Alembert, *Essai sur les eléments de philosophie*, in *Mélanges de littérature, d'histoire et de philosophie*, vol.4 (Amsterdam, Zacharis Chatelain et fils, 1759), p.27.

77. See Johann Eller, 'Seconde dissertation sur les élémens', in *Histoire de l'Académie royale des sciences et belles lettres* (1746; Berlin, Haude, 1748), p.25-48 (31): 'In order to be able to decide among the very different available opinions, it is necessary to

touchstone is a metaphorical expression for a means of testing the value of things or concepts: it represents in itself the 'trial by fire', the bottom line of experimental science able also to detect the real from the apparent – no longer a rational demonstration, but a verification. This metaphor implies a modification of the meaning of the word 'examination', which becomes more and more ***prüfen***, ***probare***, 'to put something to the test' to understand if it is made of good or bad alloy, and less the Leibnizian endless 'analysis' of concepts.[78] Lambert refers to the touchstone explicitly in his 'Discours de réception':

> If he wishes to avoid the reproach which is commonly made against philosophers, of being attached only to abstract and useless speculations, it is better for him to bring them closer to the practice and, consequently, to the experience and the individual circumstances of nature which will serve him [the speculative philosopher] as a *touchstone*, in so far as it is necessary to let him see and feel the correctness and the use which can be made of them.[79]

We can conclude that natural philosophy and mathematics are both called upon to offer metaphysics their testing power. For instance, Euler hints at the fact that the notions of space and time, being the outcome neither of sensation nor of mathematical demonstrations, can be reached only via a detour: 'the only theoretically appropriate means is to base oneself on the laws of mechanics: they are geometrically demonstrated and confirmed by experience.'[80] The mechanical

---

resort to experiments, which are, so to speak, the touchstone for discerning what is of good alloy [*de bon aloi*] in natural philosophy.' Already in use by Leibniz and Wolff, 'touchstone' became a common expression in Kant as a criterion of truth.

78. *Essai*, *épreuve* and *expérience* are regarded as synonyms; see Gabriel Girard, *Synonymes français: leurs différentes significations, et le choix qu'il en faut faire pour parler avec justesse* (Paris, Imprimerie de la veuve d'Houry, 1736). For Lambert's arguments against Leibnizian 'analysis', see his letters, particularly the letters to Kant (3 February 1766) and to Sulzer (24 July 1764, unpublished), in Johann H. Lambert, '5 lettres de Lambert et Sulzer', Basel Universitätsbibliothek, Handschriftlicher Nachlass, L.Ia 745, p.183-204.
79. 'S'il veut éviter le reproche qu'on fait assez ordinairement aux Philosophes, de ce qu'ils ne s'attachent qu'à des spéculations abstraites et inutiles, il lui convient de les rapprocher de la pratique; et par conséquent, de l'expérience et des circonstances individuelles de la Nature, qui lui serviront comme de pierre de touche, dès qu'il s'agit d'en faire voir et sentir même la justesse, de l'usage qu'on en peut faire'; Lambert, 'Discours de réception', p.510.
80. Euler to Maupertuis (2 December 1747), in *Commercium epistolicum*, p.95. In Lichtenberg's words: 'where a body is in motion, there exists space and time';

laws become, therefore, the *touchstones* for a forthcoming metaphysical theory of nature in order to avoid, in Lambert's words, any speculative talk about chimeras, or about a world which is not the one in which we live. The guarantee offered by mathematics is indeed often indirect; for instance, according to Tonelli, in Maupertuis's *Essai de cosmologie*, 'it is precisely through dynamics that mathematics has shown its utility in metaphysics.'[81]

The uneasy alliance between mathematics and teleology

Even if we acknowledge Euler's inclusion of teleology inside the core of mathematics, we might still be really astonished to find the claim in the addendum to his *Methodus inveniendi*: 'nobody has noticed as yet how this curve [i.e. the figure which a curved elastic ribbon assumes] can be investigated by the method of *maxima* and *minima*, that is to say, by final causes [*hoc est, per causas finales*].'[82] In fact, the instigator was Maupertuis who, in the same year, 1744, read his *mémoire* to the Paris Academy, and introduced, albeit with many caveats, his 'teleological' principle in mathematics: 'I know the reluctance which many mathematicians have for the final causes applied to natural philosophy, and I approve it even to a certain point; I have to admit that their introduction is not without risk.'[83]

Indeed, claims about the non-teleological nature of mathematics can be traced all the way back to Aristotle: 'in unchangeable things this principle could not exist nor could there be a good-in-itself. This is why in mathematics nothing is proved by means of this kind of cause [*final cause*], nor is there any demonstration of this kind – "because it is better, or worse".'[84] This position was also

---

see Georg C. Lichtenberg, *Notebook D (1773-1775)*, in *Aphorisms*, reprinted as *The Waste books*, translated by R. J. Hollingdale (New York, 1990), n.54, n.52.

81. Tonelli, *Maupertuis*, p.30. On the intricate relation of dynamics, physics and metaphysics in Leibniz, see Stephen Howard, 'Why did Leibniz fail to complete his dynamics?', *British journal for the history of philosophy* 25:1 (2017), p.22-40 (27-37).
82. Euler, *Methodus inveniendi, additamentum* 1, p.246.
83. 'Je connois la repugnance que plusieurs Mathématiciens ont pour les causes finales appliquées à la Physique, et l'approuve même jusqu'à un certain point; j'avoue que ce n'est pas sans peril qu'on les introduit'; Maupertuis, 'Accord de différentes loix', p.20.
84. Aristotle, *Metaph.* III, 2 996a18, in *The Complete works of Aristotle*, ed. Jonathan Barnes, 2 vols (Princeton, NJ, 1984), vol.2, p.1574.

grounded on the sharp dichotomy between 'mathematics and necessity' on the one hand and 'finalism and freedom' on the other, rooted in Aristotelian teleology. Obviously, the *impermeability* of mathematics to finality turned out to be a kind of liability for mathematics, and the neutral assumption of the ancients changed into a fundamental accusation which was still alive in the seventeenth century. By abandoning final causes, mathematics was departing from divine wisdom and sliding into the 'brute' domain of blind necessity, *sine scopo*, subjected to 'the regime of brute rules of motion and figures'.

Granted such a dichotomy between mathematics and teleology, one might be surprised by the claim of Maupertuis: 'I will soon have responded to those who blame the use I have made of the final causes in a mathematical matter: it is precisely what is mathematical in this matter that renders more victorious the application that I have made of final causes.'[85]

Where does this statement come from? The key to unravelling this knot is the fact that the teleological assumption of the least action – that 'absolutely nothing takes place in the world in which some *maximum* or *minimum property does not shine forth*'[86] – could be displayed more easily via 'maximal and minimal curves' than via a disregarded and even dangerous metaphysical display. The mathematical path is safer; because of that, one does not even need to resort to the overused argument from design. Maupertuis acknowledges the 'danger' intrinsic to the notion of final causes, and thus restricts its use precisely to mathematics where it can be applied with clarity and be tested in physics.

In Maupertuis's works we find a crucial passage, repeated in at least two fundamental texts – the 'Accord de différentes loix de la nature' and the subsequent *Essai de cosmologie* – in which *necessity* and *freedom* become compatible: 'a blind and *necessary* mechanics follows the designs of the most enlightened and *free* intelligence.'[87] Or, as he explains a few years later: 'all things are arranged so methodically,

85. 'J'aurai bientôt répondu à ceux qui blâment l'usage que j'ai fait des causes finales dans une matière mathématique: c'est justement ce qu'il y a de mathématique dans cette matière qui rend plus victorieuse l'application que j'y ai faite des causes finales'; Maupertuis, *Essai de cosmologie*, p.xiv.
86. 'Nihil omnino in mundo contingit, in quo non maximi minimive ratio quaerpia eluceat'; Euler, *Methodus inveniendi*, *additamentum* 1, p.245.
87. 'Une méchanique aveugle & nécessaire suit les desseins de l'Intelligence la plus éclairée & la plus libre'; Maupertuis, 'Accord de différentes loix', p.21.

that a blind and necessary mathematics performs what the most enlightened and free mind prescribes.'[88] We find in this passage the core of what I defined as the hybridisation of the Berlin Academy's philosophy, totally extraneous to Spinozism. Indeed, this unique coincidence of necessity and freedom is not the fruit of a metaphysical assumption of monism, rather it is the purely contingent result of a crossroads of different paths, or, as Euler once wrote, a *duplex via*.[89]

After being banned from science (*in primis* from classical mechanics: gravitational attraction could explain the dynamics of bodies without referring to any final causes), teleology re-emerges with the support of mathematics and natural philosophy in the works of two outstanding members of the Berlin Academy, Maupertuis and Euler. After them, Lagrange definitively closed the door to teleology (and to metaphysics) in science and used Maupertuis's principle as the basis of the variational calculus. It is noteworthy that even D'Alembert, the public enemy of final causes in the *Encyclopédie*,[90] nevertheless had some good words for those causes, precisely when he refers to Maupertuis's Principle of least action. When the question becomes not to seek God in the unimportant things, but 'chercher en grand', D'Alembert could turn a blind eye to the final causes, so 'dangerous' in all *a priori* considerations, but 'helpful' if stated *a posteriori*, that is, 'according to clear and incontestable principles of mechanics'.[91]

It is highly likely that in this case it was precisely the new audience and the peculiar context in which he found himself (i.e. the Berlin Academy) that induced Maupertuis to 'vindicate Newton's physics with a Leibnizian metaphysical argument'.[92]

88. 'C'est que toutes choses soient tellement ordonnées, qu'une Mathématique aveugle et nécessaire exécute ce que l'intelligence la plus éclairée et la plus libre prescrivoit'; Maupertuis, *Essai de cosmologie*, p.25.
89. Euler, *Methodus inveniendi, additamentum* 1, p.245.
90. See the entry 'Causes finales (*metaph.*)', under the entry 'Cause' in the *Encyclopédie, ou Dictionnaire raisonné des sciences, des arts et des métiers*, ed. Denis Diderot and Jean D'Alembert, vol.2 (Paris, Briasson *et al.*, 1751), p.789.
91. D'Alembert says, referring explicitly to Maupertuis: 'But if it is dangerous to make use *a priori* of final causes in order to find the laws of phenomena; it may be useful, and it is at least curious, to show how the principle of final causes agrees with the laws of phenomena, provided that these laws were first determined by clear and incontestable principles of mechanics. This is what M. de Maupertuis has proposed to do with respect to refraction in particular, in a *Mémoire* printed among those of the Academy of Sciences in 1744'; D'Alembert, *Encyclopédie*, vol.2, p.789.
92. Terrall, *The Man who flattened the earth*, p.173.

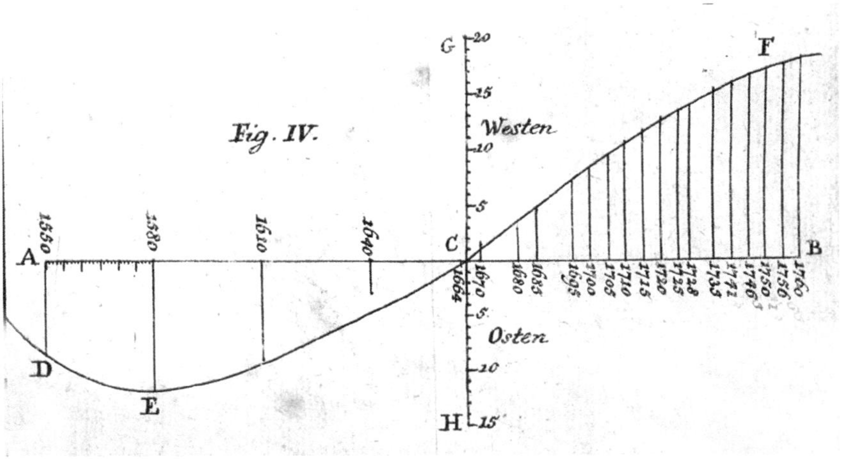

Figure 2: Figure IV from Lambert's *Theorie der Zuverlasigkeit.*

To give a visual example of such a hybrid plan, Figure 2 illustrates what 'experimental metaphysics' might look like, in order to stress the attitude towards the experimental and the speculative. The graph represents the harmonic variation of the magnetic needle over an interval of 210 years.

As we can see, the figure represents what Lambert calls a *curva manu ducta* (i.e. a curve drawn freehand), in so far as its mathematical law is unknown, and therefore constructed by just interpolating empirical data. Yet something happens at the point corresponding to the year 1640 (and even 1670): 'Above all, the ordinate for year 1640 falls behind. I think it is better to consider this observation as *uncertain* [*ungewiss*], rather than have to give an abnormal twist [*anomalistische Wendung*] to line EC.'[93] Lambert does not act in an underhand manner: with no qualms, he declares that it is worthier to consider as 'uncertain' an empirical observation – also thanks to his awareness of the difficulty of data collection, in this case – than to have anomalies in the curve. In this case, the properties of a geometrical curve (for instance: the uniformity of the variation of magnetism) are more exact and certain than empirical observations. Here is why Lambert 'reasonably supposes (§66) an adjusted curve to be more representative of the empirical data than a polynomial

93. Johann H. Lambert, *Theorie der Zuverläßigkeit der Beobachtungen und Versuche*, in *Beyträge zum Gebrauche der Mathematik und deren Anwendung* (Berlin, Buchladen der Realschule, 1792), §63, p.477.

curve drawn through every given point'.[94] And this is something intrinsically and wonderfully metaphysical.

As already noted, even an empiricist like Maupertuis, notwithstanding his aversion to the application of mathematics in philosophy, is not allowed to accept, as philosophical truths, propositions which contradict mathematical truths. The speculative plane is always lying in wait. These hybrid curves (partly empirical and partly metaphysical) illustrate the daring and steady matching of theory with data at the Berlin Academy:[95] not just an empirical recasting, but the endeavour to find, although strictly *a posteriori*, rules and laws or 'first principles' of order, both intrinsic and teleological.

Conclusions

The old sterile and blind metaphysics had clearly been dethroned and dismissed, but a new one, less insular, one whose truths should be consistent with mathematical and experimental truths, became increasingly topical in the Academy. These two turning points often coexisted with an appearance of contradiction. For instance, despite the title of his *mémoire*, 'Essai d'une démonstration *métaphysique* du principe général de l'équilibre', Euler could write: '[the question of the current movement] would doubtlessly require deeper research, which would no longer be the responsibility of metaphysics.'[96] Inevitably, 'deeper research' was no longer synonymous with metaphysics; the need for a 'higher science' offering first principles and first causes and an explanation of the nature of things, however, proved resilient. Indeed, expressions such as 'a metaphysically proven principle' were used by both Euler and Maupertuis to invoke a proof that referred to final causes (i.e. concerning questions of *maxima* and *minima*). After all, at least until the middle of the eighteenth century, the lack of

94. Oscar B. Sheynin, 'J. H. Lambert's work on probability', *Archive for history of exact sciences* 7:3 (1971), p.244-56 (254).
95. 'By claiming that scientists and philosophers should start from empirical observations, but process the data thus obtained according to axiomatic mechanisms, Lambert can be said to have anticipated the era of computer models and predictive algorithms'; Basso, 'Lambert on experience and deduction', p.194.
96. 'La question du mouvement actuel [...] ce qui demanderoit sans doute une recherche plus profonde, et qui ne sera plus du ressort de la *Métaphysique*'; Leonhard Euler, 'Essai d'une démonstration métaphysique du principe général de l'équilibre', in *Histoire de l'Académie royale des sciences et belles lettres* (1751; Berlin, Haude and Spener, 1753), p.246-54 (247).

metaphysical subtlety continued to be seen as a defect in Germany, and thereafter, while 'metaphysician' became almost an offensive term, 'metaphysics' began to be reshaped under what has been called 'the deflationary method of the experimental tradition'.[97]

We have quoted Lambert's claim: 'so here is [speculative] philosophy, closely related to experimental physics', Holland's neologism of 'experimental metaphysics' concerning Lambert's procedure, as well as Maupertuis's plain reference to 'metaphysical experiments' and Euler's hope for a new 'grand science', the 'genuine metaphysics' which includes physics and mathematics. In short, despite their different backgrounds, all share an affirmation of faith in the mathematical description of nature, in experimental proofs and in teleology. Having settled the unusual cooperation of these three features, we end up with an 'experimental metaphysics', whose characteristics come now to the fore.

Considering the odd link between experimentalism and teleology, teleology and mathematics, mathematics and experimentalism, we can look at Lambert's *Cosmological letters* and read: 'From *teleology* you derive the purposes, and from *experience* you find that the means for such purposes are available [...] You move both closer together and your *demonstration* thereby becomes *complete*.'[98] William Herschel at one point described these letters as 'full of the most fantastic imaginations';[99] accordingly, Lambert's false supposition on the existence of a satellite of Venus has been registered as 'the last *empirical* result to which his *speculative* courage led him'.[100] The question is: which class does this work belong to? Experimental philosophy, the class to which Lambert devoted himself? The mathematical class, due to its astronomical

97. Andrew Cooper, 'Kant and experimental philosophy', *British journal for the history of philosophy* 25:1 (2017), p.265-86 (275).
98. Lambert, *Cosmological letters*, p.92 (my emphasis).
99. William Herschel, 'On the direction and velocity of the motion of the sun, and solar system', *Philosophical transactions* 95 (1805), p.233-56 (235).
100. Hans Blumenberg, *The Genesis of the Copernican world*, translated by R. Wallace (Cambridge, MA, 1987), p.526 (my emphasis). In his essay 'Vom Trabanten der Venus' (in *Astronomisches Jahrbuch*, Berlin, 1777, p.186-91), Lambert predicted that Venus would also have a satellite, and that people would see it when Venus passed close to the sun, on 1 June 1777, but it was not observed. In Lambert's case, Frederick's joke fits perfectly: 'our geometrician of Berlin is in excellent health, and rather lives in the planet Venus than on this terraqueous globe'; Frederick II to D'Alembert (2 July 1769), in Frederick II, *Letters between Frederick II and M. D'Alembert*, in *Posthumous works of Frederick II*, translated by Thomas Holcroft, vol.11 (London, G. G. J. and J. Robinson, 1789), p.127.

concerns? Or speculative philosophy? As never before, it becomes evident that such dichotomies would never enclose this very hybrid domain, perfectly covered by this new 'experimental metaphysics', conceived of in opposition both to the blind 'unalloyed metaphysician' (the ***purus putus metaphysicus*** once discarded by Lambert) and to the natural philosopher who has to halt too soon.

We can find traces of this eclectic *forma mentis* among the members of the Academy,[101] and, therefore, we can conclude that something like the Principle of least action, a *metaphysical* principle, arising in *natural philosophy*, *mathematically* confirmed and then *generally* applied to everything, could be perfected only at the Berlin Academy. When scientists or mathematicians, finding themselves unwittingly in a philosophical arena, embrace their tools, namely, experiments and mathematics, in the service of speculative philosophy, they could end up building a 'Copernican revolution', 'otherwise you do not need a class of metaphysics in an Academy'.[102]

At the end of the above-quoted chapter of Maupertuis's *Lettre sur le progrès des sciences*, entitled 'Metaphysical experiments', a very inspired and allusive prophecy for metaphysics begins: 'After so many centuries, during which, notwithstanding the efforts of the greatest men, our *metaphysical knowledge* has not made the least progress, it is to be supposed that, if it is in Nature that metaphysics can make some progress, this can only be achieved by new and extraordinary tools like these.'[103]

Metaphysical 'progress' is possible only by turning to 'new tools', that is to say – in the context of this *Letter*, which calls to 'these natural philosophers who will perhaps instruct us better' – to new scientific experiments and instruments even in metaphysics.[104]

101. To quote Maupertuis: 'the art of conducting the most delicate *experiments* and that of deriving the most *ingenious consequences*' were an attitude shared by the majority of the members of the Academy ('Des devoirs de l'académicien', p.118, my emphasis; see the appendix to this volume).

102. Prémontval, 'De la psychocratie: fin', p.416.

103. 'Après tant de siecles écoulés, pendant lesquels, malgré les efforts des plus grands hommes, nos connoissances métaphysiques n'ont pas fait le moindre progrès, il est à croire que s'il est dans la nature qu'elles en puissent faire quelqu'un, ce ne sauroit être que par des moyens nouveaux, & aussi extraordinaires'; Pierre-Louis M. de Maupertuis, *Lettre sur le progrès des sciences*, in *Œuvres de Maupertuis*, vol.2, p.373-431 (430).

104. This challenge may have been subsequently taken up by Kant, who, in order to place metaphysics in 'the secure course of a science', applied 'Bacon's

The age of a universal method was over. The idea of synergy among disciplines, which we now face, breaks completely from the traditional one, which hinged on the method. Natural philosophy and mathematics, as shown, are called upon as *touchstones* to offer to metaphysics their testing power and results, and no longer as methodological models.[105]

The result of the entanglement between metaphysics, mathematics and the empirical sciences is not a hotchpotch, but, as Euler perfectly explained, the setting for a sort of handover: one starts as soon as the other ceases, following the same track. What is new here is that metaphysics is no longer the starting point, the queen of the sciences which delivers first principles to all the sciences, but the end; as Merian wrote: 'It may be said in general that each part of natural philosophy reaches a point beyond which a new field for the speculative philosopher [*le spéculateur*] opens up.'[106] Whenever we are extended beyond our experience, metaphysics is within reach. As Lambert confessed in his *Cosmological letters*: 'one will grant me the possibility and finally even agree that the conclusions from my system are *not contrary to experience.* This is also all that I can at most demand.'[107] In the context of *maxima* and *minima*, the 'experimental level' of this peculiar 'experimental metaphysics' spans from a *minimum* where it is merely 'not contrary to experience' to a *maximum* where it overlaps with natural philosophy.

---

experimental method to the practice of metaphysics' (Cooper, 'Kant and experimental philosophy', p.266).

105. Dealing with the contribution of mathematics in metaphysics through dynamics (in relation to the proof of God's existence), Maupertuis in his *Essai de cosmologie* shows how 'this is a particular application of mathematics to metaphysics, and not a metaphysics *more geometrico* demonstrated' (Tonelli, *Maupertuis*, p.30). Lambert himself, despite being committed to the Cartesian and Wolffian ideal of a universal *mos geometricus* (seen as nonsense by Maupertuis and Euler), in his endless endeavours to answer the Academy's prize question on evidence, ended up with his *New Organon*, an insight into new instrumental philosophical sciences: 'Un ouvrage fort long et qui a du rapport avec la question philosophique que Votre Illustre Académie de Berlin ira couronner'; Lambert to Euler (7 March 1763), in *L. Euler und J. H. Lambert Briefwechsel*, ed. Karl Bopp (Berlin, 1924), p.32-33.

106. 'On peut dire en général que chaque partie de la Physique aboutit à un terme où s'ouvre un nouveau champ pour le spéculateur'; Merian, 'Discours sur la métaphysique', p.455.

107. Lambert, *Cosmological letters*, p.95 (my emphasis).

Origins of the arts, origins of man in Sulzer's academic essays

ALESSANDRO NANNINI

The essays which Johann Georg Sulzer read from 1751 in the sittings of the class of speculative philosophy of the Académie royale des sciences et des belles-lettres in Berlin have been examined in recent years from several points of view, with particular regard to psychology and aesthetics.[1] The major interest of commentators has perhaps been the dichotomy between feeling and knowing, explicitly theorised by Sulzer in his 'Observations sur les divers états où l'âme se trouve en exerçant ses facultés primitives, celle d'apercevoir et celle de sentir' of 1763 ('Observations on the different states in which the soul finds itself while using its primitive faculties, that of knowing and that of feeling'),[2]

This work was supported by a grant of the Ministry of Research, Innovation and Digitization, CNCS/CCCDI – UEFISCDI (project number PCE 105/2021, within PNCDI III).

1. For Sulzer's relationship with the Academy, see Johan Van der Zande, 'Johann Georg Sulzer – Spaziergänge im Berliner Tuskulum', in *Berliner Aufklärung: Kulturwissenschaftliche Studien*, ed. Ursula Goldenbaum and Alexander Košenina, vol.1 (Hanover, 1999), p.41-68 (50-55). A recent issue of the journal *Philosophiques* 42:1 (2015), ed. Christian Leduc and Daniel Dumouchel, was devoted to philosophical research at the Berlin Academy. Two essays directly concern Sulzer: Daniel Dumouchel, 'Entre métaphysique et esthétique: sentiment et beauté dans les mémoires de Johann Georg Sulzer', *Philosophiques* 42:1 (2015), p.51-72, and Stefanie Buchenau, '"L'œil ne se voit pas voir": Sulzer sur la contemplation et le sentiment de soi', *Philosophiques* 42:1 (2015), p.73-88. Both of them, in different ways, acknowledge the link between psychology and aesthetics. See also the essays in *Johann Georg Sulzer (1720-1779): Aufklärung zwischen Christian Wolff und David Hume*, ed. Frank Grunert and Gideon Stiening (Berlin, 2011), and *Johann Georg Sulzer: Aufklärung im Umbruch*, ed. Elisabeth Décultot, Philipp Kampa and Jana Kittelmann (Berlin, 2018).
2. Johann Georg Sulzer, 'Observations sur les divers états où l'âme se trouve en exerçant ses facultés primitives, celle d'apercevoir et celle de sentir', in *Histoire de*

which has traditionally been interpreted as a departure from Wolffian foundations towards the autonomy of sentiment.[3] Although the most recent investigations have reduced the emphasis on this departure,[4] the problem of sensibility in general remains at the heart of the historiographical debate, in particular in its link with the energetic conception of the fine arts,[5] which grounds the monumental project of the *Allgemeine Theorie der Schönen Künste* (*General theory of the fine arts*, 1771-1774).

While sensibility is crucial in countless respects in Sulzer's work, from psychology to aesthetics, the present chapter focuses in particular on its role in anthropological issues. More specifically, I aim to analyse the relevance of sensibility in Sulzer's account of the

l'Académie royale des sciences et belles lettres (1763; Berlin, Haude and Spener, 1770), p.407-20.

3. Wolfgang Riedel, 'Erkennen und Empfinden: Anthropologische Achsendrehung und Wende zur Ästhetik bei Johann Georg Sulzer', in *Der ganze Mensch: Anthropologie und Literatur im 18. Jahrhundert*, ed. Hans-Jürgen Schings (Stuttgart, 1994), p.410-39; Caroline Torra-Mattenklott, *Metaphorologie der Rührung: Ästhetische Theorie und Mechanik im 18. Jahrhundert* (Munich, 2002), p.227-94; Achim Vesper, 'Le plaisir du beau chez Leibniz, Wolff, Sulzer, Mendelssohn et Kant', *Revue germanique internationale* 4 (2006), p.23-36; Elisabeth Décultot, 'Von der Seelenkunde zur Kunsttheorie: zu Sulzers Untersuchung über den Ursprung der angenehmen und unangenehmen Empfindungen (1751/52)', *Scientia poetica* 12 (2008), p.69-88; Elisabeth Décultot, 'Die Schattenseiten der Seele: zu Johann Georg Sulzers Theorie der dunklen Vorstellungen', in *Formen des Nichtwissens der Aufklärung*, ed. Hans Adler and Rainer Godel (Paderborn, 2010), p.263-78. This autonomisation of sentiment is said to have influenced Kant: Anton Palme, *Sulzers Psychologie und die Anfänge der Dreivermögenslehre* (Berlin, 1905), p.3, 36-43 and 75-77; Rita Terras, 'The development of the doctrine of the three faculties in German aesthetics', *Monatshefte* 70:1 (1978), p.5-14; Marion Heinz, 'J. G. Sulzer und die Anfänge der Dreivermögenslehre bei Kant', in *Johann Georg Sulzer (1720-1779)*, ed. F. Grunert and G. Stiening, p.83-100.

4. Werner Euler, 'Die Idee des Schönen in Sulzers allgemeiner Theorie des Vergnügens', in *Johann Georg Sulzer (1720-1779)*, ed. F. Grunert and G. Stiening, p.101-33 (115-16); Gideon Stiening, 'Zur physischen Anthropologie einer Unsterblichkeit der Seele', in *Johann Georg Sulzer (1720-1779)*, ed. F. Grunert and G. Stiening, p.57-81 (p.63, n.42; p.64, n.54); Dumouchel, 'Entre métaphysique et esthétique'.

5. *L'Esthétique de Johann Georg Sulzer (1720-1779)*, ed. Bernard Deloche (Lyon, 2005); Ernst Stöckmann, 'Von der sinnlichen Erkenntnis zur Psychologie der Emotionen', in *Physis und Norm: neue Perspektiven der Anthropologie im 18. Jahrhundert*, ed. Manfred Beetz, Jörn Garber and Heinz Thoma (Göttingen, 2007), p.69-106; Ernst Stöckmann, *Anthropologische Ästhetik: Philosophie, Psychologie und ästhetische Theorie der Emotionen im Diskurs der Aufklärung* (Tübingen, 2009).

awakening of humans from their original slumber, drawing on all his relevant essays delivered at the Berlin Academy until his death. A first attempt in this direction was made in 1969 by Walter Moser, who studied the influence of French sensualist thought (especially Condillac) on the Berlin Academy, but did not develop Sulzer's doctrine in all its articulations.[6] What remained on the periphery of Sulzer research was the investigation of the cultural dimension of that awakening, in so far as it concerns not only individuals, but also whole peoples.

It has only been since the renewed focus on Sulzer's essay about the origins of the fine arts, the *Pensées sur l'origine et les différens emplois des sciences et des beaux-arts* (*Thoughts about the origin and the different uses of the sciences and the fine arts*), read at the Berlin Academy on 27 January 1757 and published in a stand-alone volume in the same year, that scholarship has laid the foundations for a more theoretically aware treatment of the latter issue.[7] In light of this development, I intend to examine the relationship between the individual and the social aspect in the emergence of the human across several of Sulzer's essays, enhancing the essential mediation of sensibility in both these domains. Such a comparative analysis is precisely what is needed for a fresh and more comprehensive examination of the role aesthetics plays in the 'anthropopoietic' process for the Swiss thinker, promising at the same time to make a significant contribution to the unearthing of the radical assumptions and implications of what has been labelled as 'aesthetic education' in the late German Enlightenment.[8] In the first section of this chapter, I examine Sulzer's theses about the issue of the soul, with particular regard to its initial condition, while in the second section I investigate the role of the body in its awakening. In

6. Walter Moser, 'Jean-Georges Sulzer, continuateur de la pensée sensualiste dans l'Académie de Berlin', *Modern language notes* 84 (1969), p.931-41.
7. Johann Georg Sulzer, *Pensées sur l'origine et les différens emplois des sciences et des beaux-arts* (Berlin, Haude and Spener, 1757). See Matthew Riley, *Musical listening in the German Enlightenment: attention, wonder and astonishment* (Aldershot, 2004); Elisabeth Décultot, 'Johann Georg Sulzers System der schönen Künste', in *Johann Georg Sulzer (1720-1779)*, ed. F. Grunert and G. Stiening, p.211-24; Dumouchel, 'Entre métaphysique et esthétique', §6.
8. See David Pugh, 'Aesthetic humanism and its foes: the perspective from Halle', in *Who is this Schiller now? Essays on his reception and significance*, ed. Jeffrey L. High, Nicholas Martin and Norbert Oellers (Rochester, NY, 2011), p.116-32; Alessandro Nannini, 'Umanizzare l'uomo: le implicazioni antropologiche dell'educazione estetica nell'epoca della *Popularphilosophie*', *Cultura tedesca* 50 (2016), p.101-17.

the third section, I advocate that the original state of deadly slumber is typical not only of the individual soul, but also of primitive peoples, pinpointing in both cases the relevance of *aisthesis* in the way out of passivity. If the process of humanisation thus relies upon aesthetic stimulations, I argue, it nonetheless implies an equally important level of anaesthetisation, whereby social imperatives are engraved in the soul, turning humans into citizens. My conclusion is that according to Sulzer it is precisely through this dialectical tension between *aisthesis* and *anaisthesis*, which the birth of the arts in each people brings to the fore, that humans begin to exist as such, in a process which never comes to a final end.

The soul as an active force

The linchpin around which Sulzer's academic studies are organised is undoubtedly the notion of soul. We get to this notion, prior to philosophical analysis, by way of introspection. Sulzer writes in his late work, 'Observations sur quelques propriétés de l'âme comparées à celles de la matière: pour servir à l'examen du matérialisme' ('Observations on some properties of the soul compared to those of matter: an examination of materialism', 1771):

> A reflexive attention paid to what happens in us can convince us that we are active beings. We feel that there is a force or an energy in us which continually tends to produce some changes both within and without. This activity seems to be the most essential property of our being, which is in itself sufficient to account for all our affections.[9]

As early as the first part (1751) of the four 'Recherches sur l'origine des sentimens agréables et désagréables' ('Studies on the

9. Johann Georg Sulzer, 'Observations sur quelques propriétés de l'âme comparées à celles de la matière: pour servir à l'examen du matérialisme', in *Nouveaux mémoires de l'Académie royale des sciences et belles-lettres de Berlin* (1771; Berlin, Voss, 1773), p.390-410 (391): 'Une attention réflèchie à ce qui se passe en nous peut nous convaincre que nous sommes des êtres actifs. Nous sentons qu'il y a en nous une force ou énergie qui tend continuellement à produire quelque changement soit au-dedans, soit au dehors. Cette activité paroît même être la propriété la plus essentielle de notre être, qui seule suffit pour rendre raison de toutes nos affections.' All translations in this chapter are my own unless otherwise indicated. Original texts have not been modernised for spelling or punctuation.

origin of pleasant and unpleasant feelings'), the Swiss philosopher had suggested a strategy to prove the existence of this dimension:

> In order to know the nature of this force, we just need to represent it in very remarkable situations, for example in a strong passion. Everybody knows that we are pressed and shaken by the force of desire. In the cases in which the soul is quieter, the essential force remains the same, albeit weaker; it always elicits an agitation similar to that of the passions, whether it is stronger or weaker. This is what the phrase 'essential force of the soul' means.[10]

As is affirmed at the beginning of the essay on apperception (1753), however, the empirical method cannot be severed from a 'metaphysical theory of man' or 'physics of the soul',[11] which rationally grounds its results. Provided that 'the essential aspect of

10. Johann Georg Sulzer, 'Recherches sur l'origine des sentimens agréables et désagréables, première partie: théorie générale du plaisir', in *Histoire de l'Académie royale des sciences et belles lettres* (1751; Berlin, Haude and Spener, 1753), p.57-75 (64): 'Pour bien connoitre la nature de cette force, nous n'avons qu'à nous la représenter dans des cas fort remarquables, comme p. e. dans une grande passion. Tout le monde sait combien on est alors pressé & troublé par la force du desir. Dans les autres cas où l'ame est plus tranquille, la force essentielle ne laisse pas d'être la même, quoique moins grande; elle excite toujours une agitation semblable à celle des passions, plus ou moins forte. Voilà ce que veut dire le terme de *force essentielle de l'ame.*'
11. Johann Georg Sulzer, 'Sur l'apperception et son influence sur nos jugemens', in *Histoire de l'Académie royale des sciences et belles lettres* (read in 1753, published in the proceedings for 1764; Berlin, Haude and Spener, 1766), p.415-34 (415); Johann Georg Sulzer, 'Analyse du génie', in *Histoire de l'Académie royale des sciences et belles lettres* (1757; Berlin, Haude and Spener, 1759), p.392-404 (392); Johann Georg Sulzer, 'Explication d'un paradoxe psychologique: que non seulement l'homme agit et juge quelquefois sans motifs et sans raisons apparentes, mais même malgré des motifs pressans et des raisons convainquantes', in *Histoire de l'Académie royale des sciences et belles lettres* (1759; Berlin, Haude and Spener, 1766), p.433-50 (434); Johann Georg Sulzer, *Kurzer Begriff aller Wißenschaften und andern Theile der Gelehrsamkeit, worin jeder nach seinem Inhalt, Nutzen und Vollkommenheit kürzlich beschrieben wird* (1745), 2nd edn (Frankfurt and Leipzig, Langenheim, 1759), p.156. See Wolfgang Proß, '"Meine einzige Absicht ist, etwas mehr Licht über die Physik der Seele zu verbreiten": Johann Georg Sulzer (1720-1779)', in *Helvetien und Deutschland: Kulturelle Beziehungen zwischen der Schweiz und Deutschland in der Zeit von 1770-1830*, ed. Hellmut Thomke, Martin Bircher and Wolfgang Proß (Amsterdam, 1994), p.134-48; Maurizio Pirro, 'Sulzers Physik der Seele und die Dramentheorie Schillers', *Zeitschrift für Germanistik* 16 (2006), p.314-23; Dumouchel, 'Entre métaphysique et esthétique'; Stefanie Buchenau, 'Sulzers

human nature is constantly the same at all times and in all climates',[12] the soul can be construed as a substance, whose natural activity consists in thinking, that is, in producing and comparing ideas. The equivalence between the activity of the soul or its 'essential force' and the production of ideas, Sulzer adds, has already been solidly established by the 'illustrious' Wolff,[13] even though, as made clear in his 'Analyse du génie' ('Analysis of genius', 1757), the source of this doctrine is the 'great' Leibniz: 'All the faculties of the soul arise from that original force which, according to the great *Leibniz*, constitutes the essence of each substance, and in particular of the soul. This active force, more easily felt in one's gut than defined, is the one which produces all ideas, or at least which compels us incessantly to develop and multiply them.'[14]

Sulzer's conclusion is of consequence: 'The essence of our soul, the principle from which all our constant desires arise, is a powerful determination to produce and receive ideas.'[15] In this sense, Sulzer argues in the essay on apperception, there is an evident analogy between the soul and the permanent forces which are present in matter, for example, the force of reaction, which does not depend on a disposition of parts (as in the case of the elastic force), does not get exhausted if it is triggered one thousand times, and cannot be considered as non-existent if it never manifests.[16] On the basis of this analogy, Sulzer feels entitled to maintain that the soul is not material.[17] Given the premises, the conclusion could sound paradoxical; and yet, if it is true that the same properties can be attributed both to the soul and to the permanent forces of matter, it is also true that in this latter case we have to do with passive forces of resistance, whereas the soul is an absolutely active and

---

Physik der Seele zwischen Medizin und Philosophie', in *Johann Georg Sulzer: Aufklärung im Umbruch*, ed. E. Décultot, P. Kampa and J. Kittelmann, p.36-50.

12. Sulzer, 'Recherches [...] première partie: théorie générale du plaisir', p.60.
13. Sulzer, 'Recherches [...] première partie: théorie générale du plaisir', p.61.
14. Sulzer, 'Analyse du génie', p.394: 'Toutes les facultés de l'ame tirent leur origine de cette force primitive, qui, selon la remarque du grand *Leibnitz*, fait l'essence de toute substance, & particulièrement celle de l'ame. C'est cette force active, qu'il est plus facile de sentir au-dedans de soi-même, que de définir, qui produit toutes les idées, ou qui du moins nous engage continuellement à les développer & à les multiplier'; see Stöckmann, *Anthropologische Ästhetik*, p.216-20.
15. Sulzer, 'Recherches [...] première partie: théorie générale du plaisir', p.63.
16. Sulzer, 'Sur l'apperception', p.419.
17. Sulzer, 'Sur l'apperception', p.419-20.

spontaneous force. Such a force cannot be removed from the soul, not even by annihilation of the latter. It follows, therefore, that it exists regardless of the body: 'The whole destruction of the organised body will plunge the soul into a mortal sleep, stifling any of its activities, without its forces ceasing to exist.'[18]

The example of the reaction force permits Sulzer to make another crucial point: in a mass of matter, there are forces which are absolutely dead until another mass acts on them; only then does matter become active because of its resistance and of the change which it provokes on the other mass.[19] With suitable precautions, this statement also applies to the force of the soul.[20] The force of the soul thus finds itself in a condition of striving by its own nature, which, however, it cannot satisfy in isolation. Its determination to produce ideas expresses itself, as it were, in a succession of empty states, which condemns it to exist in a dormant condition: 'Sentiments do not arise with man, no more than passions.'[21] In this sense, Aristotelians err in advocating the comparison of the soul with a slate covered in wax or a blank sheet, but are totally right in arguing that 'all our perceptions originally come from the senses'.[22] Hence, in order for the pending state of the soul to be interrupted, something must impact on our sensory organs: 'The soul would be but a dead force, which would remain in an eternal inaction. To make it act, it is absolutely necessary that it be solicited from without.'[23] If the body is not a necessary requirement for the soul's existence, nevertheless, its mediation becomes essential for passing from a condition of mere latency, in which the inner activity runs on idle, to a condition of self-consciousness.

18. Sulzer, 'Sur l'apperception', p.420.
19. Sulzer, 'Sur l'apperception', p.418-19.
20. In the essay on materialism, Sulzer writes: 'The elastic force acts in the same moment and in an opposite direction to the action of the strike, which consequently cannot produce it; [the strike] just gives it the occasion to manifest itself. The same holds true for our inner activity, which does not deploy itself when we lack perceptions'; Sulzer, 'Observations sur quelques propriétés', p.392-93.
21. Johann Georg Sulzer, 'Recherches sur l'origine des sentimens agréables et désagréables, quatrième partie: des plaisirs moraux', in *Histoire de l'Académie royale des sciences et belles lettres* (1752; Berlin, Haude and Spener, 1754), p.373-90 (386).
22. Sulzer, 'Observations sur quelques propriétés', p.395.
23. Sulzer, 'Sur l'apperception', p.418.

The role of the body

Sulzer plainly acknowledges the importance of the body for the awakening of the soul: 'It seems that we are obliged to admit as a matter of fact that the soul does not feel anything, does not know anything and does not have any clear perception, not even of its own existence, if not through the ministry of the body.'[24] By virtue of the body, the soul comes into contact with other forces whereby it can 'awaken and recognise itself'.[25] If self-consciousness is 'the act of the spirit through which, distinguishing our being from the ideas occupying us, we distinctively know what we are doing and what is happening in us and around us',[26] the presence of the body conveys the proximity of the external world which elicits the primordial act of discrimination between our soul and known objects.[27] In this way, the 'intercession' of the body enables the soul to relinquish the condition of near-death – Sulzer speaks of 'sommeil mortel'[28] or 'assoupissement mortel' – in which the soul 'despite its immortality, [would be] a lifeless being'.[29]

On his account, the relation between the body and the soul relies more on a plausible hypothesis than on a certain demonstration, as is explicitly admitted in the third *mémoire* on the immortality of the soul (1776): 'I confess that I have neither facts nor reasonings to prove that the soul cannot feel anything without the aid of the body.'[30] Yet Sulzer can count on what he considers to be an unshakeable axiom:

24. Johann Georg Sulzer, 'Sur l'immortalité de l'âme considérée physiquement: premier et second mémoire', in *Nouveaux mémoires de l'Académie royale des sciences et belles-lettres* (1775; Berlin, Voss, 1777), p.359-87 (359): 'Il semble en effet qu'on est obligé d'admettre comme une chose de fait que l'ame ne sent rien, ne connoît rien & n'a aucune perception claire, pas même celle de sa propre existence, que par le ministere du corps.'
25. Johann Georg Sulzer, 'Sur l'immortalité de l'âme considérée physiquement: troisième mémoire', in *Nouveaux mémoires de l'Académie royale des sciences et belles-lettres* (1776; Berlin, Voss, 1779), p.349-59 (352).
26. Sulzer, 'Sur l'apperception', p.416; see Udo Thiel, 'Sulzer über Bewusstsein im Kontext', in *Johann Georg Sulzer (1720-1779)*, ed. F. Grunert and G. Stiening, p.21-36.
27. If our condition is clearer than that of known objects, we employ the faculty of feeling; if known objects are clearer than our condition, we employ the faculty of knowing. See the end of this section for further detail.
28. Sulzer, 'Sur l'apperception', p.420.
29. Sulzer, 'Sur l'immortalité de l'âme [...]: premier et second mémoire', p.359.
30. Sulzer, 'Sur l'immortalité de l'âme [...]: troisième mémoire', p.352.

> The soul is undoubtedly an active being; it is an always acting force. We know by experience that it can act without being aware of it and that it feels itself very weakly when nothing hinders its activity, whereas it feels itself with the maximum of energy when it faces obstacles which hinder it in a remarkable way; it is then that [the soul] makes efforts and strongly feels its state. Therefore, it seems likely to me that it is necessary that an external force comes up to interrupt or hinder its activity.[31]

Certainly, the 'number of forces which continuously manifest themselves in nature and act in every corner of the universe is almost endless'.[32] It is up to our body to give an order to the myriad of stimuli which incessantly strike us, preventing the exposure of the soul to an unimaginable quantity of sensations which would necessarily condemn it to a state of permanent confusion: 'The organised body could therefore moderate this overabundance of sensations, intercepting the majority of these sensations and, through sensory organs, only allowing those sensations which comply with the general order of things of this universe to get to the soul.'[33]

In this way, it is clear that the body, as well as permitting the soul to act on the material world, has the function of conveying the impressions coming from the latter.[34] Consequently, it also concentrates such impressions according to a certain order, of which the senses are the fundamental channels. This is why 'the eye seems to have the function of a convex glass which gathers together an endless number of luminous rays in a very small room, in order to render their impulses stronger; and the ear has the same effect on

31. Sulzer, 'Sur l'immortalité de l'âme [...]: troisième mémoire', p.352: 'L'ame est indubitablement un être actif; c'est une force toujours agissante. Nous savons par l'expérience qu'elle peut agir sans le savoir & que même elle se sent très foiblement, lorsque rien ne la gêne dans son activité; au lieu qu'elle se sent avec le plus d'énergie lorsqu'elle rencontre quelque obstacle qui la gêne un peu fortement; c'est alors qu'elle fait des efforts & qu'elle sent vivement son état. Il me paroît donc probable qu'il faut qu'une force extérieure vienne, ou l'interrompre, ou la gêner dans son activité, pour qu'elle s'éveille & se reconnoisse.'
32. Sulzer, 'Sur l'immortalité de l'âme [...]: troisième mémoire', p.353.
33. Sulzer, 'Sur l'immortalité de l'âme [...]: troisième mémoire', p.353: 'Le corps organisé pourroit donc modérer cette surabondance de sensations en interceptant le plus grand nombre de ces sensations & en ne faisant parvenir à l'ame, moyennant les organes, que celles qui conviennent à l'ordre général des choses de cet Univers.'
34. Sulzer, 'Sur l'immortalité de l'âme [...]: premier et second mémoire', p.365.

the acoustic rays.'[35] The conclusion is not long in coming: 'It seems, therefore, that one of the main functions of the animal body is to receive and modify the impressions of the material world, so as to render them perceivable to the soul, which would perhaps not be touched at all without this instrument, or it would be [touched] in a completely different way.'[36]

If it is the soul which perceives in the strict sense, Sulzer does not shy away from the debate, very lively at the end of the eighteenth century, on its possible location.[37] In particular, he clearly indicates the brain as its adequate seat, by virtue of the mediation of an 'animated molecule',[38] at the same time material and immaterial. It is a subtle body, able to escape from the death to which the animal body is doomed, which has the property of 'feeling' on the basis of the impressions transmitted by the sensory nerves:

> It is possible to prove with incontestable facts that the faculty of feeling or, to explain myself in a more positive sense, the being which feels, is not spread or disseminated for all the extent of the animal body, but has its seat in the brain. It is not the eye which sees or the ear which hears. The soul feels the impression of the light and sound only when, by means of the nerves, the movement produced in the organs is conveyed to the cerebral medulla. [...] This incontestably proves that perceptions are produced only within the brain and that the sensory organs serve only to receive the impressions of the objects and to modify them, so that they can get to the seat of the soul.[39]

35. Sulzer, 'Sur l'immortalité de l'âme [...]: premier et second mémoire', p.365.
36. Sulzer, 'Sur l'immortalité de l'âme [...]: premier et second mémoire', p.366: 'Il paroît donc qu'une des fonctions principales du corps animal est de recevoir & de modifier les impressions du monde matériel de manière à les rendre sensibles à l'ame, qui sans cet instrument n'en seroit peut-être point affectée du tout, ou le seroit tout différemment.'
37. See Stiening, 'Zur physischen Anthropologie'; *Das Gehirn: Organ der Seele? Zur Ideengeschichte der Neurobiologie*, ed. Ernst Florey and Olaf Breidbach (Berlin, 1993).
38. See Stiening, 'Zur physischen Anthropologie', p.76; see also Daniela Gay, 'L'amitié entre Johann Georg Sulzer (1720-1779) et Charles Bonnet (1720-1793)', in *Les Ecrivains suisses alémaniques et la culture francophone au XVIIIe siècle*, ed. Michèle Crogiez Labarthe, Sandrine Battistini and Karl Kürtös (Geneva, 2008), p.85-96.
39. Sulzer, 'Sur l'immortalité de l'âme [...]: premier et second mémoire', p.363: 'D'abord, on peut prouver par des faits incontestables que la faculté de sentir, ou pour m'expliquer plus positivement, que l'être qui sent n'est pas répandu ou disséminé par toute l'étendue du corps animal, mais qu'il a son siege

To the explanation of this transmission, Sulzer had already dedicated a good part of his essay on sensible pleasures of 1752: 'We know by experience', he asserted in this *mémoire*, 'that of all the changes happening in the universe, our soul can only feel those which cause certain impressions on the sensory organs.'[40] It is not important for Sulzer to state whether this law is a necessary consequence of the nature of the soul, as seems likely, or an arbitrary disposition of God; what matters here is that the only way to know its changes of nature is via the senses. 'But', the Swiss philosopher wonders, 'what does it mean to feel?'[41] Our experience only shows that we employ this word in correspondence with strong and lively perceptions of certain things which seem to be their physical causes in us. Without entering into a discussion about the metaphysical hypothesis at the basis of this relation – a discussion that was at the core of Wolffian rational psychology – Sulzer limits himself to averring that '*the impressions of the sensory organs excite and produce the sensations in the soul*, without having the pretence of or professing or disagreeing with any of these systems [influxionism, occasionalism and armonicism] conceived to explain the union of the soul with the body.'[42]

Despite his caution, Sulzer's elucidation decidedly inclines towards influxionism.[43] The starting point adopts an attitude of prudence:

> In the impressions received by the senses, it is necessary to seek the cause or, if you prefer, the analogy or the occasion of the sensations

---

dans le cerveau. Ce n'est pas l'œil qui voit, ni l'oreille qui entend. L'ame ne sent les impressions de la lumiere & de sons que lorsque moyennant les nerfs le mouvement produit dans les organes est communiqué à la moëlle du cerveau [...]. Cela prouve incontestablement que les perceptions ne se font que dans l'intérieur du cerveau, & que les organes des sens ne servent qu'à recevoir les impressions des objets, pour les modifier de manière qu'elles puissent parvenir au siege de l'ame.'

40. Johann Georg Sulzer, 'Recherches sur l'origine des sentimens agréables et désagréables, troisième partie: des plaisirs des sens', in *Histoire de l'Académie royale des sciences et belles lettres* (1752; Berlin, Haude and Spener, 1754), p.350-72 (352).
41. Sulzer, 'Recherches [...] troisième partie: des plaisirs des sens', p.352.
42. Sulzer, 'Recherches [...] troisième partie: des plaisirs des sens', p.353: '*les impressions des organes des sens excitent ou produisent les sensations dans l'ame*, sans prétendre, ni adopter, ni choquer, aucun de ces systèmes inventés pour expliquer l'union de l'ame avec le corps' (original emphasis).
43. Falk Wunderlich, 'Johann Georg Sulzers Widerlegung des Materialismus und die Materietheorien der Zeit', in *Johann Georg Sulzer (1720-1779)*, ed. F. Grunert and G. Stiening, p.37-55 (52-54).

> of the soul. I therefore contend that every sensation is caused by some movement of the nerves of the body, and I pose as a principle: *That the soul does not have sensations in the absence of an analogous movement in the nerves of the senses.*[44]

Such an analogy must be understood in the sense of a proportion on the one hand between the strength (or vividness) of the sensation of the soul and the quantity of movement in the nerves, and on the other hand between the variation of the movement and the variation of the sensation. Sensibility and the condition of the nerves will consequently determine the differences of the senses, whereby, Sulzer argues, 'the essence of the senses in general consists in the nerves.'[45] Only the nerves which are proportionate to the relevant stimulus can be activated by the latter; hence, the eye is also hit by acoustic rays, though without producing any perceivable outcome.

In light of this doctrine, the quantity of sensation, that is, its vividness, relies not only on the quality of the nerves (with greater power assigned to the 'coarser' senses: in the first place touch, and then taste, smell, hearing and sight),[46] but also on the quantity of movement in the nerves, which is in turn linked with the quantity and velocity of colliding matter.[47] Hence, what seems to be an uninterrupted perception stems from a succession of discrete impacts, which are too rapid to be singularly perceived and capable of letting the nerves vibrate for the time necessary for the production of the perception in the soul.[48] As has been highlighted by Proß,[49] and then also pinpointed by Dürbeck[50] and Torra-Mattenklott,[51] the model of such a neurological argumentation is the Halle physiological

44. Sulzer, 'Recherches [...] troisième partie: des plaisirs des sens', p.353: '[I]l faut chercher dans les impressions, que les sens recoivent, la cause, ou si l'on veut l'analogie, ou l'occasion des sensations de l'ame. Je dis donc, que toute sensation est causée par quelque mouvement des nerfs du corps, & je pose pour principe; *Que l'ame n'a point de sensation, sans un mouvement analogue dans les nerfs des sens*' (original emphasis).
45. Sulzer, 'Recherches [...] troisième partie: des plaisirs des sens', p.353.
46. Sulzer, 'Recherches [...] troisième partie: des plaisirs des sens', p.358 and 360.
47. Sulzer, 'Recherches [...] troisième partie: des plaisirs des sens', p.355.
48. Sulzer, 'Recherches [...] troisième partie: des plaisirs des sens', p.355-56.
49. See Proß, 'Meine einzige Absicht', p.143-48, who also points to Daniel Georg Coschwitz's studies among Sulzer's sources.
50. Gabriele Dürbeck, *Einbildungskraft und Aufklärung: Perspektiven der Philosophie, Anthropologie und Ästhetik um 1750* (Tübingen, 1998), p.134.
51. Torra-Mattenklott, *Metaphorologie der Rührung*, p.227-94 (239).

school. It is, in particular, Krüger's *Naturlehre* which supplies the relevant theoretical bases to surmise the proportionality between movement and sensation,[52] as well as to outline a tentative hypothesis to calculate the strength of the sensation,[53] thereby developing some hints already present in Wolff's *Psychologia rationalis* (*Rational psychology*).[54]

In the wake of this line of thought, Sulzer affirms that every total sensation is made up of a great number of simple sensations, provoked by a succession of impressions of the same strength as those at the basis of a uniform sound.[55] If it is true that we can define 'momentum of sensation' as the perception which the soul feels following an immediate impression, Sulzer continues, all the physiological considerations expounded can be traced back to the momentum of sensation and to the sensitivity of the nerves.[56] In addition, the sensations deriving from the succession of different momenta should be taken into account, relying on the rapidity of the succession and their variety.[57] What is important to notice in this complex articulation, intertwining psychological motifs with more genuinely neurological issues, is the progressive opacification of the corporeal dimension, which is not reducible to a mere external 'interface' of the force of the soul. Sulzer maintains in the 'Analyse de la raison' ('Analysis of reason', 1758): 'We do not have any measurement of this force [...]. Yet, we know that it is hindered in its activity by the current state of the organs of the body, so that it develops itself only in proportion to this state.'[58]

52. Johann Gottlieb Krüger, *Grundriß eines neuen Lehrgebäudes der Artzneygelahrheit* (Halle, Hemmerde, 1745), p.49; Johann Gottlieb Krüger, *Naturlehre*, 2nd edn, 3 vols (Halle, Hemmerde, 1744-1755), vol.2, p.83. The theoretical bases for this connection are already present in Descartes and other seventeenth-century philosophers. On this aspect in Descartes, see, for example, Ann Scholl, *Descartes's dreams: imagination in the Meditations* (New York, 2005), ch.1: 'Theory of sensory perception'. On this aspect in Locke, see, for example, Martha Brandt Bolton, 'Locke on sensory representation', in *Perception and reality: from Descartes to the present*, ed. R. Schumacher (Paderborn, 2004), p.146-67.
53. Krüger, *Naturlehre*, vol.2, p.585-88.
54. Christian Wolff, *Psychologia rationalis* (Frankfurt and Leipzig, Renger, 1734), §118.
55. Sulzer, 'Recherches [...] troisième partie: des plaisirs des sens', p.356.
56. Sulzer, 'Recherches [...] troisième partie: des plaisirs des sens', p.357.
57. Sulzer, 'Recherches [...] troisième partie: des plaisirs des sens', p.361.
58. Johann Georg Sulzer, 'Analyse de la raison', in *Histoire de l'Académie royale des sciences et belles lettres* (1758; Berlin, Haude and Spener, 1765), p.414-42 (416).

In so far as the sensory organs order and concentrate external stimuli, conveying them to the brain according to the mentioned rules, the different quality of the nerves will determine – all stimuli being equal – different psychological effects. Thus, a man with a vigorous temperament has more lively perceptions than another, because the strength of the sensation has become for him the natural way of feeling. It follows that the reason why the activity of the soul turns to certain stimuli and not to others, as well as their intensity, also depends on the way in which the outer forces are elaborated by corporeal complexion: 'The question of how the constitution of the body can alter the action of the soul', Sulzer significantly comments, 'is one of the most embarrassing'.[59] The only ground of certainty in this respect is that our inner feeling must necessarily derive, albeit in a manner for us inexplicable, from the concurrence of the active force of the soul and of a certain organisation of the body.[60] This means, on the one hand, that the active force of the soul can be endowed with greater or lesser intensity – as is made clear in the 'Analyse du génie' (1757), it is the souls of geniuses which perceive the strongest desire, and are thereby more attentive and receptive with respect to what touches the senses and imagination.[61] On the other hand, the nature of the active force notably varies in relation to the nature of the body:

> It is likely that [this active force of the soul] depends to a great degree on the constitution of the body. [...] The soul acts and feels only in proportion to the impressions which it receives from the body, and does react very weakly if there is a dulling of the senses. The liveliest man can plunge into a state of stupidity if some weakness takes possession of his senses. The active force of the soul, therefore, depends on natural causes which concur to form the temperament and the constitution of the body. Birth, climate and diet contribute to it most remarkably.[62]

59. Sulzer, 'Analyse de la raison', p.416.
60. Sulzer, 'Analyse de la raison', p.416-17.
61. Sulzer, 'Analyse du génie', p.394-95. That the soul takes an active interest in what it receives is proof against materialism; see Sulzer, 'Observations sur quelques propriétés', p.393. This aspect is particularly emphasised by Wunderlich, 'Sulzers Widerlegung des Materialismus', p.43-45. As explained in his first essay on pleasure, contingencies are crucial for the direction taken by the soul in the course of one's life; see Sulzer, 'Recherches [...] première partie: théorie générale du plaisir', p.71.
62. Sulzer, 'Analyse du génie', p.402: '[I]l est probable qu'elle dépend, en grande partie, de la constitution du corps. [...] De même l'ame ne sent & n'agit que

Hence, if it is true that the barren existence of the essential force of the soul is independent from the body, its possibility to express itself and the way in which it expresses itself are intimately linked to its corporeal basis, a basis which is tangible both in the capacity of gathering stimuli and in the quality of their transmission. It is in this way that the body can direct the natural inclinations of the soul towards certain perceptions rather than towards others, actively forging our tastes and idiosyncrasies:

> The specific taste stemming from the active force of the soul is still substantially determined by the constitution of the body. It is by way of the body that the soul has a relationship with the universe, and it receives the knowledge of objects only from the organs of the body. We therefore understand that the constitution of these organs necessarily influences the active force of the soul.[63]

If Baumgarten had already clearly stated that we represent the universe according to the position of our body,[64] this thesis now gains fresh momentum, which helps clarify the awakening of the soul from its original slumber. For in this process the intensity of the stimuli is mediated by the presence of a body which, far from being a transparent vector of the outer world, betrays its temperamental and neurological density, and is able to affect the preferences whereby the soul tends towards objects.

We can therefore conclude that the initial awakening stems precisely from a considerable and repeated variation of the inner force, due to a series of solicitations to which the natural conformation

---

proportionnellement aux impressions qu'elle reçoit du corps, & elle n'agit que fort foiblement si les sens sont émoussés. L'homme le plus vif peut tomber dans un état de stupidité si quelque foiblesse s'empare de ses sens. La force active de l'ame dépend donc beaucoup des causes naturelles qui concourent à former le tempérament & la constitution du corps. La naissance, le climat, & la nourriture y contribuent le plus.'

63. Sulzer, 'Analyse du génie', p.402: 'Le goût spécifique qui nait de la force active de l'ame est encore en grande partie déterminé par la constitution du corps. C'est par le corps que l'ame tient à l'Univers, & elle ne reçoit la connoissance des objets que par ses organes. De là on comprend que la constitution de ces organes influe nécessairement sur la force active de l'ame.' Along with physical causes, moral causes, such as education and exercise, should be taken into account in the development of certain inclinations (see p.402-403). For this issue, see below.
64. Alexander Gottlieb Baumgarten, *Metaphysics* (1st edn 1739; 4th edn 1757), ed. and translated by Courtney D. Fugate and John Hymers (London, 2013), §513.

of the body itself is particularly sensitive, thus having an immediate relationship with the soul. Thanks to the effect of such stimuli and to the subsequent variation of their inner activity, humans for the first time fasten their attention upon themselves. As stated in the essay of 1763, when attention has a target lying outside ourselves, the soul exercises the faculty of knowing; when attention turns inwards, the soul feels, and in particular it feels itself.[65] Depending on the kind of effect provoked by the sensory stimulus, therefore, the soul starts to feel its being in a pleasant or unpleasant way,[66] hence through an emotionally laden experience.

If this 'auroral' shock in preconscious flatness is present, albeit no longer retrievable by way of introspection, in the history of each individual, something similar also occurs in the history of peoples. As Sulzer explicitly contends in the 'Considérations psychologiques sur l'homme moral' ('Psychological considerations on the moral man') of 1769, 'what happens to the individual happens to the whole species', or 'to a whole people'.[67] It is precisely the examination of the collective plane that gives Sulzer the possibility to prove his psychological theory within a wider historical frame. Because of this shift of emphasis, the point will be to understand what kind of sensory solicitation originally brings each group of human-beasts to their fully fledged humanity.

65. Johann Georg Sulzer, 'Observations sur les divers états', p.410. When we employ the faculty of feeling, the ideas of the objects provoking the sentiment are confused or obscure (p.410-11). As noted above, this is a much-debated issue which deserves an in-depth investigation in its own right.

66. At the basis of pleasure lies the possibility for the soul to be fostered in its activity, while displeasure implies some sort of obstacle or hindrance in its activity; see Sulzer, 'Recherches […] première partie: théorie générale du plaisir'. See also Sulzer, 'Observations sur quelques propriétés', p.391. The problem of pleasure in Sulzer has been widely studied recently; see for example Elisabeth Décultot, 'Métaphysique ou physiologie du beau? La théorie des plaisirs de Johann Georg Sulzer (1751-1752)', *Revue germanique internationale* 4 (2006), p.93-106; Lorenzo Lattanzi, 'Il dibattito della *Aufklärung* sull'oscurità del sentimento gradevole', in *Il secolo dei lumi e l'oscuro*, ed. Piero Giordanetti, Giambattista Gori and Maddalena Mazzocut-Mis (Milan, 2009), p.161-205 (201). On the links with Condillac, see Moser, 'Jean-Georges Sulzer'.

67. Johann Georg Sulzer, 'Considérations psychologiques sur l'homme moral', in *Histoire de l'Académie royale des sciences et belles lettres* (1769; Berlin, Voss, 1771), p.361-80 (368-69).

Savages and humans

The initial situation of these anthropomorphic beings is marked by a bestial life, where physiological needs can be satisfied only through instincts. As Sulzer writes in the essay on moral pleasures: '[The savage] sees the sky and nature, but no interesting idea rises out of them. He remains in a stupidity and in an insensibility which is on a par with that of the brutes.'[68] Stupidity and insensibility, the lack of knowledge and feeling, point to a total passivity in relation both to inner life and to external objects.[69] It is this condition that represents the cultural equivalent of what was the liminal dimension of a soul not yet hit by a stimulus capable of affecting it. In this stage, savages perceive pleasure as a mere effect of necessity,[70] which does not leave any trace on the surface of their soul, since the link between the satisfaction of needs and self-preservation is not yet understood:[71] 'Half-savage peoples, as those native Americans who are not capable of counting to four, can remain seated for a whole day without thinking, just repeating the same note on their flute thousands of times without getting bored.'[72]

Primary needs, however, enable individuals gradually to concentrate their attention upon objects important for survival, distinguishing for example water from other materials.[73] Such needs are at the basis of the mechanical arts, which 'all concern the production and the preparation of the things crucial to the needs and comfort of life'.[74] With the development of the mechanical arts, hence with the

68. Sulzer, 'Recherches [...] quatrième partie: des plaisirs moraux', p.389.
69. In this sense, Riley (*Musical listening*, p.68) is completely right in arguing that 'the cultivation of the attention – and thus the releasing of the human-beast from its prison of constant astonishment – is a central task of the civilizing process.' In fact, the human-beast is unable to know and to feel, in so far as it is unable to turn its attention inwards and outwards, being incessantly at the mercy of the most violent stimuli.
70. Sulzer, *Pensées sur l'origine*, p.12; Sulzer, 'Considérations psychologiques', p.366.
71. Sulzer, 'Considérations psychologiques', p.366. The infinitely small degree of virtue which is present in this stage is only the effect of instinct; see Sulzer, *Pensées sur l'origine*, p.12.
72. Johann Georg Sulzer, *Allgemeine Theorie der Schönen Künste* (1771-1774), 4 vols (Leipzig, Weidmanns Erben und Reich, 1778-1779), vol.3, p.211.
73. Johann Georg Sulzer, 'Observations sur l'influence réciproque de la raison sur le langage et du langage sur la raison', in *Histoire de l'Académie royale des sciences et belles lettres* (1767; Berlin, Haude and Spener, 1769), p.413-38 (417).
74. Sulzer, *Kurzer Begriff aller Wißenschaften*, §69.

entrance into society,[75] half-savage peoples progressively multiply the number of their interactions, consequently increasing the activity of the soul as well as the need to keep it engaged:

> As soon as men widen their horizon through social life and increase their inner activity, the absence of thought which characterises them – since neither the spirit nor the feelings are stimulated and revived by external objects – becomes unbearable. Only the half-savage man, who has hardly arisen out of the animal, can bear such an absence of thought; but if it is often experienced by a man in part already more cultivated, this latter loses his activity and vitality of his spirit and heart, which properly render him a being by far superior to animals.[76]

While it is primary needs which concentrate the attention on the objects capable of bringing satisfaction, it is nature which originally turns attention on oneself thanks to the variety of pleasant sensations it produces. Such an element is particularly evident in the *Pensées sur l'origine et les différens emplois des sciences et des beaux-arts* (1757). Although sciences and the arts stem from the same origin – the human genius – and tend to the same objective – to overcome natural barbarity – the different faculties which they address produce a crucial divergence in their evolution. Relying on speculation, sciences attain the acme of their development only after progressing for centuries, in accordance with the slowness of advancement of reasoning, whereas the fine arts, hinging on the immediacy of sentiment, reach their highest perfection soon after their invention.[77]

75. On the state of nature in Sulzer, see Johann Georg Sulzer, 'Recherches sur un principe fixe, qui serve à distinguer les devoirs de la morale de ceux du droit naturel', in *Histoire de l'Académie royale des sciences et belles lettres* (1756; Berlin, Haude and Spener, 1758), p.450-56 (452).
76. Sulzer, *Allgemeine Theorie*, vol.4, p.109: 'Sobald die Menschen durch das gesellschaftliche Leben ihren Gesichtskreis erweitert, und ihre innere Würksamkeit vermehrt haben, wird ihnen der gedankenlose Zustand, da weder der Geist noch die Empfindung durch äußere Gegenstände gereizt und in einige Wärme gesetzt werden, unerträglich. Nur der noch halb wilde Mensch, der sich wenig über das Thiere empor gehoben hat, kann einen solchen Zustand der Gedankenlosigkeit ertragen: stellt er sich aber bey dem schon etwas mehr gebildeten Menschen oft ein, so verlieret dieser dadurch seine Würksamkeit und die Wärme des Geistes und Herzens, die ihn eigentlich zu einem weit über die Thiere erhabenen Wesen machen.' A similar conception is found in Condillac; for more details, see Moser, 'Jean-Georges Sulzer'.
77. Sulzer, *Pensées sur l'origine*, p.7-9.

What is the origin of the fine arts? As a saying taken up by Sulzer goes, the arts arise from indigence, hence, in line with Charles Batteux,[78] the mechanical arts precede the fine arts, which aim at pleasure and at the engagement of the imagination and heart.[79] In order for the fine arts to originate, therefore, a certain independence from primary needs is required, which enables one to widen one's horizon beyond mere material self-preservation:

> As soon as man found himself in the condition of satisfying with ease his more pressing needs, he had the opportunity to reflect on needs less necessary to his mere survival, and yet essential to a comfortable and pleasant existence. He then perceived that the objects capable of pleasantly influencing his senses, imagination and heart were not present in a great abundance in nature nor were they always in his power.[80]

This is the moment when the fine arts come into play. As is immediately clear, there is a close connection between the birth of the arts and the birth of humanity. Both of them rise from the richness in sensations offered by an environment with a mild climate. Natural pleasures leave a strong impression on the most sensitive souls in each people, whose instincts spontaneously tend to sweet sensations. It is these 'génies assez heureux' who awaken first from lethargy and are more inclined to gather a certain number of pleasant objects, copy them and intensify their effect, so as not to leave the cause of these delightful sensations at the mercy of climate and seasonal contingencies.[81]

78. The argument is famously stated in Charles Batteux, *The Fine arts reduced to a single principle* (1746), translated by James O. Young (Oxford, 2015), p.3.
79. Sulzer, *Kurzer Begriff aller Wißenschaften*, §69; see Décultot, 'Johann Georg Sulzers System der schönen Künste', p.215-16. On the origins of the arts in the eighteenth century, with special reference to French authors, see Fernando Bollino, *Ragione e sentimento: idee estetiche nel Settecento francese* (Bologna, 1991); Christian Michel, 'L'argument des origines dans les théories des arts en France à l'époque des Lumières', in *Primitivisme et mythes des origines dans la France des Lumières 1680-1820*, ed. Christian Michel and Chantal Grell (Paris, 1989), p.35-45.
80. Sulzer, *Pensées sur l'origine*, p.14-15: 'Dès que l'homme se trouva en état de satisfaire avec facilité aux besoins les plus pressans, il eut le loisir de faire des reflexions sur des besoins moins necessaires à sa simple conservation, mais aussi essentiels à une existence aisée & agréable. Il sentit dès-lors, que les objets capables d'affecter agréablement ses sens, son imagination & son cœur, n'étoient pas en grande abondance dans la nature, ni toujours en son pouvoir.'
81. Sulzer, *Pensées sur l'origine*, p.14-15.

Although these instincts ensure the presence of the fine arts in each people, the first works of art are produced in casual experiments aimed at increasing the pleasant sensations perceived in correspondence with certain objects. Through these works, the earliest artists manage to elicit an agreeable variation in the activity of the soul of their fellow humans, thus serving as true mediators of humanity. But in what way would the fine arts be able to drag savages out of the stifling embrace of catalepsy?

In his essay 'De l'énergie dans les ouvrages des beaux-arts' ('On energy in works of fine arts') of 1765, written under Henry Home's influence,[82] Sulzer seems to provide the bases for a possible answer. As argued two years earlier, when a man exercises his faculty of feeling, his soul enters into a state of sentiment[83] or emotion[84] in which his attention turns to himself more than to the external object striking him. Sulzer divides the causes of emotion into three classes, attributable, in ascending order of importance: (a) to an interruption of a uniform flux of perceptions by means of new and unexpected objects (comparable to the recovering of consciousness while waking up);[85] (b) to the 'energetic' liveliness of the representations of objects in themselves already pleasant or unpleasant,[86] which prompts a series of transitions between contemplation and emotion;[87] and (c) to the capacity of immediately affecting passions.[88] It is these forces which provide the sensible object with the best weapons to strike the soul,[89]

82. See Annie Lamblin, 'Johann Georg Sulzer', in *Aux sources de l'esthétique: les débuts de l'esthétique philosophique en Allemagne*, ed. Jean-François Goubet and Gérard Raulet (Paris, 2005), p.283-94 (288); Sven Gesse, *Genera mixta: Studien zur Poetik der Gattungsmischung zwischen Aufklärung und Klassik-Romantik* (Würzburg, 1997), p.63-64.

83. Sulzer, 'Observations sur les divers états', p.410.

84. Johann Georg Sulzer, 'De l'énergie dans les ouvrages des beaux-arts', in *Histoire de l'Académie royale des sciences et belles lettres* (1765; Berlin, Haude and Spener, 1767), p.475-92 (476).

85. Sulzer, 'De l'énergie dans les ouvrages des beaux-arts', p.477-79.

86. See Michel Delon, *L'Idée d'énergie au tournant des Lumières: 1770-1820* (Paris, 1988), p.107.

87. Sulzer, 'De l'énergie dans les ouvrages des beaux-arts', p.479-85; on contemplation, see Buchenau, 'L'œil ne se voit pas voir'.

88. Sulzer, 'De l'énergie dans les ouvrages des beaux-arts', p.485. See the entry 'Kraft' in the *Allgemeine Theorie* for a different interpretation of this doctrine.

89. See Bernard Deloche, 'Johann Georg Sulzer, théoricien de l'effet esthétique', in *L'Esthétique de Johann Georg Sulzer*, ed. B. Deloche, p.39-67.

thus complementing the subjective premises examined in the first part of this chapter.

Not by chance, Sulzer cursorily mentions these three forms of aesthetic energy in the passage of the *Pensées* of 1757 in which he describes the beginning of humanity: initially, savages perceive the breaking presence of pleasant sensations from sensory objects dispersed in nature, which interrupt the state of slumber of their soul. When the environment stimulating them is no longer present, natural objects, which are particularly pleasant, can be re-presented to the imagination[90] through 'a lively painting either by means of words or by means of drawing and colours'.[91] Moreover, humans can by now not only slavishly copy objects, but also imitate them more freely, embellishing them and making them richer than the originals themselves. In this way, it is easier to engage, along with the senses and imagination, also the heart and passions: 'In place of the too uniform birdsong, they imagined a more melodious, varied song, supported by harmony and able to remarkably flatter the passions, of which it imitated the tone, movement and expression. This is the origin of the fine arts.'[92] As is apparent from the last statement, music plays a particular role in this genetic process.[93] Admittedly, hearing enjoys a unique position among the senses: on the one hand, it is one of only two senses, together with sight, which makes it possible to know the cause of the sentiments it triggers, thus presenting beautiful objects to us; on the other hand, it is coarser than sight, so its sensations are stronger. In fact, hearing is the most powerful sense addressed by the fine arts.[94]

90. In the essay on intellectual pleasures, imagination is defined as a supplement of the senses; see Johann Georg Sulzer, 'Recherches sur l'origine des sentimens agréables et désagréables, seconde partie: théorie des plaisirs intellectuels', in *Histoire de l'Académie royale des sciences et belles lettres* (1751; Berlin, Haude and Spener, 1753), p.76-100 (79).
91. Sulzer, *Pensées sur l'origine*, p.15.
92. Sulzer, *Pensées sur l'origine*, p.16: 'Au lieu du chant trop uniforme des oiseaux, on imagina un chant plus melodieux, plus varié, soutenû par l'harmonie & capable de flatter sensiblement les passions, dont il imitoit le ton, le mouvement & l'expression. Voilà l'origine des Beaux-Arts.'
93. See Riley, *Musical listening*, p.65-78; Jean-Joël Duhot, 'Sulzer et l'esthétique musicale', in *L'Esthétique de Johann Georg Sulzer*, ed. B. Deloche, p.103-38; Thomas Christensen, 'Introduction', in *Aesthetics and the art of musical composition in the German Enlightenment*, ed. Nancy K. Baker and Thomas Christensen (Cambridge, 1995), p.3-24.
94. See Sulzer, 'Recherches [...] seconde partie: théorie des plaisirs intellectuels', p.78; see also Sulzer, *Allgemeine Theorie*, vol.3, p.75.

If Dubos believed that the eye is nearer to the soul than the ear,[95] Sulzer responds that there is an absolutely direct connection between hearing and heart.[96]

It is not overstating things, therefore, to conclude that music is the first art to take over the baton from nature in awakening humans from their original inertia.[97] The figure of Orpheus for the Greeks is here exemplary and makes clear what counts as a general anthropological truth:[98] 'Certainly, it is not entirely legendary', Sulzer writes on the one hand, 'what the Greek tradition tells us about Orpheus, who must have dragged the Greeks out of their wild state through music. What other means could be used to bring a savage people to attention and to sentiment?'[99] On the other hand, relying on Horace's teachings, Sulzer remarks that 'Orpheus tamed raw and cruel men through the enchantment of his lyre, in order to lead them to their duties.'[100]

While, in the first case, men take the place of the stones and plants of the original myth, which the aesthetic force of Orpheus' music is able to vivify and charm, in the second case, the emphasis lies on the moral aspect of their awakening. In fact, the savage is not only passive, but also, as recalled in the *Pensées*, raw and cruel; the transition, as it were, from *hominitas* to *humanitas* requires mitigating this rawness, by stirring up more delicate sentiments through sweeter and continuous stimuli: 'Through pleasant sensations flowing to us from all sides, nature undoubtedly wanted in general to form our minds to mildness and sensibility [...]. It is in this way that we become humans; our activity increases, because the things which we

95. Jean-Baptiste Dubos, *Réflexions critiques sur la poésie et sur la peinture* (1719; Paris, 1993), p.133.
96. Sulzer, *Allgemeine Theorie*, vol.3, p.264.
97. Sulzer will clearly argue in his *Allgemeine Theorie* that rhythmisation arises out of instinct; see Clémence Couturier-Heinrich, '"Tendance naturelle au rythme" et "observation instinctive de la mesure": l'inscription du rythme dans le domaine du spontané chez Johann Georg Sulzer et August Wilhelm Schlegel', *Revue germanique internationale* 18 (2002), p.53-70.
98. On the role of Orpheus in Renaissance and early modern rhetoric, see Heinrich Franz Plett, *Rhetorik der Affekte: Englische Wirkungsästhetik im Zeitalter der Renaissance* (Tübingen, 1975), p.176-81. For the importance of this figure in Sulzer's philosophy, see Johan Van der Zande, 'Orpheus in Berlin: a reappraisal of Johann Georg Sulzer's theory of polite arts', *Central European history* 28 (1995), p.175-208.
99. Sulzer, *Allgemeine Theorie*, vol.3, p.275-76.
100. Sulzer, *Pensées sur l'origine*, p.29.

find interesting increase [...]; and we rise from the dust towards the nobility of superior beings.'[101]

If, therefore, the taming of cruelty, which can 'appease man's natural fierceness',[102] is a crucial aspect of humanising in so far as it prepares him to take care of his duties, how is it possible to bring people to effectively observe such duties, especially the natural duty to perfect oneself? In the 'Considérations', it is incontestable that man comes to know his responsibilities through reflection, whereby the savage begins to perceive the necessity of actions that he has performed by instinct up until then.[103] As explained in the 'Analyse de la raison', reflection entails the faculty of commanding one's attention which is the basis of reason.[104] Savagery, however, is not tamed just because of an abstract knowledge of duties: one must also put them into practice.

As already suggested in the essay on apperception[105] and then more plainly in the essay on psychological paradoxes,[106] reason is too weak a motive to influence man's judgements and guide his actions; hence, it is not through a mere rational conviction that humans can be led to their duties. The example of prejudices, though, seems to offer a different solution. Sulzer writes:

> There are thousands of things which we heard countless times before the age of reflection. We have become so acquainted with these judgements that every time one of these things comes back, the relevant judgement returns at the same time, and this happens so rapidly that every perception seems to be an inner sensation. When it

101. Sulzer, *Allgemeine Theorie*, vol.3, p.57: 'Ohne Zweifel wollte die Natur durch die von allen Seiten auf uns zuströmenden Annehmlichkeiten unsre Gemüther überhaupt zu der Sanftmuth und Empfindsamkeit bilden [...]. Dadurch werden wir zu Menschen; unsre Thätigkeit wird vermehret, weil wir mehrere Dinge interessant finden [...]; wir heben uns aus dem Staub empor, und nähern uns dem Adel höherer Wesen.' The English term 'sensibility' covers the semantic range of the German terms *Sinnlichkeit* and *Empfindsamkeit*, both of which are used by Sulzer in the *Allgemeine Theorie*: while the first term refers to knowledge through the senses (internal or external), the second refers to a particular delicacy in receiving certain stimuli. For a semantic explanation of this issue, see Johann Georg Sulzer, *Teoria generale delle Belle Arti* (1771-1774), partial Italian edition by Alessandro Nannini (Bologna, 2011).
102. Sulzer, *Pensées sur l'origine*, p.26.
103. Sulzer, 'Considérations psychologiques', p.366.
104. Sulzer, 'Analyse de la raison', p.422.
105. Sulzer, 'Sur l'apperception', p.426.
106. Sulzer, 'Explication d'un paradoxe psychologique'.

comes to acting, the obscure motives, consequences of these so-called sensations, cannot fail to have their effect, whatever our reason says.[107]

The point is therefore to give to rational truths the immediacy and efficacy of the obscure motives: 'What we have just observed about the ease whereby we receive the most bizarre opinions and practices and about the obstinacy whereby we stick to them shows us that it would not be difficult to inspire reasonable opinions and sentiments in a naive and unbiased people, and so to make them wise and virtuous.'[108] On the one hand, it is necessary to set positive models in the education both of a whole people and of the young. On the other hand, it is essential to endorse the frequent repetition of a given reasoning, so as to render it more and more familiar, until it is assimilated in the mass of obscure ideas. In this way, these elements are reflectively considered no longer as ideas external to us, but as modifications of our own being, instantaneously affecting our preferences and tastes.[109] Just as the natural body provides temperamental and neurological bonds to our way of feeling, 'incorporated' truths[110] provide us with ethical bonds, which are no less compelling. Indeed, to obey one's duties would here mean nothing other than to obey the inclinations of our second nature.

107. Sulzer, 'Explication d'un paradoxe psychologique', p.446: 'Il y a mille choses sur lesquelles nous avons entendu prononcer une infinité de fois, avant l'âge de réflexion. Ces jugemens nous sont devenus si familiers, que toutes les fois qu'une de ces choses-là nous revient, le jugement que nous en avons entendu porter revient en même tems, & cela se passe si rapidement, que toute la perception paroit être une sensation intérieure. Lors donc qu'il est question d'agir, les motifs obscurs, conséquences de ces prétendues sensations, ne peuvent pas manquer d'avoir leur effet, quoiqu'en dise la raison.'

108. Sulzer, 'Explication d'un paradoxe psychologique', p.449: 'Ce que nous venons d'observer sur la facilité avec laquelle on reçoit les opinions et les pratiques les plus bisarres & sur l'opiniâtreté avec laquelle on s'y attache, nous fait voir qu'il ne seroit pas difficile d'inspirer à un peuple simple & non prévenu, des opinions & des sentimens raisonnables, de le rendre sage & vertueux'; see Décultot, 'Die Schattenseiten der Seele'; Daniel Dumouchel, 'Tiefen der Seele: veränderte Zustände und psychologische Paradoxe – die empirische Psychologie bei J. G. Sulzer', in *Johann Georg Sulzer: Aufklärung im Umbruch*, ed. E. Décultot, P. Kampa and J. Kittelmann, p.14-35; Alessandro Nannini, 'At the bottom of the soul: the psychologization of the *fundus animae* from Leibniz to Sulzer', *Journal of the history of ideas* 82:1 (2021), p.51-72.

109. Sulzer, 'Considérations psychologiques', p.370.

110. Sulzer, 'Explication d'un paradoxe psychologique', p.448-49; see also Sulzer, 'Considérations psychologiques', p.369.

Beginning with the aforementioned *Pensées* (1757) and the 'Réflexions philosophiques sur l'utilité de la poésie dramatique' ('Philosophical reflections about the usefulness of dramatic poetry') of 1760, it is clear that the fine arts, as mistresses of the human heart, will be crucial for both these strategies.[111] Admittedly, if the fine arts are rightly steered by philosophy, they manage to submit to sensibility a vast array of virtuous models, gradually forging the deepest regions of the soul in the course of time: 'The arts', Sulzer argues, 'are able to take hold of man's heart and conduct.'[112] As a result, the fine arts are not only capable of presenting our duties in a fascinating way, thus instantly triggering the driving forces of the soul,[113] but, by virtue of the repeated effect of their aesthetic energy, they can also surreptitiously lay down the traces of the contents they convey. Only through this process of incorporation can the savage take a fully human appearance – an appearance on whose outlines the specific cultural mark that has redeemed him from previous ferity is engraved. It is not by chance that Orpheus is described as an expert psychagogue,[114] awakening savages as self-conscious men with the same gesture whereby he calls them to be civilised citizens.

If, therefore, our humanity is the result of a genuinely 'aisthetic' action, which makes us feel ourselves for the first time, it also entails an 'anaesthetic' side, which 'numbs' in one's soul the imperatives of a given society.[115] Humanisation thus becomes the tension *par excellence* between *aisthesis* and *anaisthesis*, between the emotional discovery of what we are and the deaf incorporation of what we have to be.[116] As

111. Sulzer, *Pensées sur l'origine*, p.28-29; Johann Georg Sulzer, 'Réflexions philosophiques sur l'utilité de la poésie dramatique', in *Histoire de l'Académie royale des sciences et belles lettres* (1760; Berlin, Haude and Spener, 1767), p.326-40. Décultot sees in the education of sensibility the connecting link between psychology and the fine arts in Sulzer; see Elisabeth Décultot, 'De la science de l'âme à la théorie des beaux-arts: l'œuvre de Johann Georg Sulzer et l'esthétique des Lumières', in *Les Lumières: un héritage et une mission – hommage à Jean Mondot*, ed. Gilbert Merlio and Nicole Pelletier (Pessac, 2012), p.177-92 (184). See also Achim Vesper, 'Sulzer über die schönen Künste und das Gute', in *Johann Georg Sulzer (1720-1779)*, ed. F. Grunert and G. Stiening, p.169-89.

112. Sulzer, *Pensées sur l'origine*, p.27.

113. Sulzer, *Pensées sur l'origine*, p.29.

114. Sulzer, *Allgemeine Theorie*, vol.3, p.62.

115. In the *Allgemeine Theorie* Sulzer will clearly state that becoming an honest citizen of Sparta, Athens or Rome are very different things; see Sulzer, *Allgemeine Theorie*, vol.2, p.43.

116. This aspect reminds us of the duplicity of sentiment, alternatively taken in a

Sulzer suggests in the *Pensées*, such a tension does not exclusively belong to the incipient moment of humankind, since people can always sink back into the inactivity of stones and the wildness of brutes without the watchful presence of some Orpheus: 'If the object capable of occupying us is missing, the spirit falls into boredom, the bodily humours are corrupted, man [...] falls into a stupid insensibility which likens him to beasts. It is the fine arts which pluck him from this company, banishing the brute indolence of the uncultivated spirit.'[117]

As clarified above, individuals need a strong stimulus impacting their senses in order to originally awaken as human beings; now it is apparent that this humanising stimulation does not resemble a long-forgotten event, which guarantees the indefinite continuation of our inner activity, but is rather something which can never cease to happen, if we are to preserve our status as humans. The same holds true for the taming of uncivilised natures,[118] which the concept of incorporation lets us see in its procedural dimension as the inexhaustible attempt to sculpt moral ideals in the ground of the soul.[119] In so far as humanity is not meant as a threshold to trespass once and for all, but as a ridge always poised between lethargy and savagery, Sulzer's argument clearly states that the fine arts can never abdicate their humanising role.[120]

In view of the multiplicity of theoretical perspectives implied by such a conclusion, from psychology and physiology to morals

---

psychological or in a moral sense; see the entry 'Empfindung' in the second volume of the *Allgemeine Theorie.*

117. Sulzer, *Pensées sur l'origine*, p.24-25: 'Si l'objet capable de nous occuper manque, l'esprit tombe dans l'ennui, les humeurs du corps se corrompent, l'homme [...] tombe dans une stupide insensibilité qui l'associe aux bêtes. Ce sont les Beaux-arts, qui le tirent de cette compagnie, en chassant l'indolence brutale naturelle à l'esprit non cultivé.'

118. Sulzer, *Pensées sur l'origine*, p.25-26.

119. Sulzer, *Pensées sur l'origine*, p.26: 'It is undoubtedly a great service paid by the fine arts for the civilization of humans to make them active and pleasant to themselves, and to arouse them to a rank which is by far superior to that of beasts.' On the controversy with Rousseau, see Johan Van der Zande, 'Thespis for and against: Sulzer and Rousseau on theater and politics', in *Reconceptualizing nature, sciences and aesthetics*, ed. Patrick Coleman (Geneva, 1998), p.231-46.

120. Not by chance, Sulzer intends to use the results of his study on the origin of the fine arts to explain their character and function; see Sulzer, *Pensées sur l'origine*, p.21. For analysis, see Dumouchel, 'Entre métaphysique et esthétique', §6.

and aesthetics, this could well be the most significant legacy that the academic essays bequeath to the aesthetic education of the late Enlightenment, as well as to Sulzer's magnum opus, the *Allgemeine Theorie der Schönen Künste.* Its frontispiece cartoon,[121] which depicts a number of savages in the background and the Graces in the foreground guarded by Athena, should, therefore, be interpreted not merely as the initial victory over passive instincts and rawness, but as a challenge which both the fine arts and political power have to take up again and again.

121. See Van der Zande, 'Orpheus in Berlin'; Carsten Zelle, 'Encyclopédisme et esthétique: la dimension européenne de Sulzer', in *Ferments d'ailleurs: transferts culturels entre Lumières et romantismes*, ed. Denis Bonnecase and François Genton (Grenoble, 2010), p.187-212.

Appendix
'On the duties of the academician'[1]

BY PIERRE-LOUIS MOREAU DE MAUPERTUIS

Gentlemen, as I undertake here to speak of the duties of the academician, I need only to speak of what you are doing, to all but say what you ought to do. And I would have given this form to my address had I not feared that it would have an air of ostentation about it, for which one could reproach me, in spite of how little a part I play in your glory and your work. I will speak here, therefore, of the duties of the academician in general. If you find it praises you, those who are not part of this body will find out what would make them worthy of it.

But before I speak of duties to free men, as are those who are citizens of the Republic of Letters, I ask, what is the law that can obligate them? Why should the philosopher renounce this liberty for which he seems to have sacrificed everything in order to subject himself to these duties? To be attached to fixed occupations of a certain kind? It is certainly necessary that he finds some advantage in it; but what is that advantage?

1. 'Des devoirs de l'académicien', in *Histoire de l'Académie royale des sciences et belles lettres, depuis son origine jusqu'à présent* (Berlin, Haude and Spener, 1750), p.115-25. Reprinted with variations in *Histoire de l'Académie royale des sciences et belles lettres, depuis son origine jusqu'à présent* (Berlin, Haude and Spener, 1752), p.137-47, *Histoire de l'Académie royale des sciences et belles lettres* (1753; Berlin, Haude and Spener, 1755), p.511-21, with the following footnote after the title: 'This address was delivered in the public assembly of 18 June 1750 and it has already been published in the volume entitled, *Histoire de l'Académie depuis son origine*. But its natural place is in the "Mémoires", and the volume from the year to which it belongs would have presented it had M. de Maupertuis not had a serious illness at that time which obscured this aim. It is to remedy this omission that we place it here, as one of the most interesting pieces that the Academy can preserve.' Also reprinted with variations in *Œuvres de Maupertuis*, 4 vols (Lyon, Bruyset, 1768), vol.3, p.283-302. A loose English translation was published in 1751 in *The Gentleman's magazine and historical chronicle* 21 (May 1751), p.218-22.

It is that which people derive from all societies; it is the mutual help that all of those who are its members lend one another. Each society possesses a common good, from which each individual member draws much more than he himself contributes.

If a man who applies himself to the sciences wants to be self-sufficient, does not want to borrow from anyone else the knowledge of which he has need, even should he have, as I will suppose, all the genius that is possible, how painful and slow will his /116/ progress be! How much time will he lose in discovering those truths that he would have already known had he profited from the help of others? He will have exhausted his strength before he even arrives at the point from where he could have begun. How much more does the man, guided by the illumination of those who came before him and of his contemporaries, save all his vigour only for those difficulties that they have not already resolved, how much more is this man in a state to solve them?

All this help that we find dispersed in the writings of and exchanges between scientists, the academician finds brought together in an Academy. He profits from them painlessly in the affability of the society, and he has the pleasure of owing them to colleagues and to friends. Add to this what is more important still: he acquires in our gatherings that *esprit académique*, that kind of feeling for the truth, which prompts him to discover it wherever it is, and to avoid looking where it is not. How many different authors have ventured systems, the falsity of which academic discussion could have revealed to them! How many chimeras would they have not dared to produce in an academy!

I have cited for you here, Gentlemen, only the immediate advantages that each academician finds in his association with an academy: it was with those advantages that I had to commence when speaking to philosophers. There are others which, while not direct means, ought to be powerful motives to excite the educated: the protection with which sovereigns honour academies, and the gifts that they shower on those who distinguish themselves among them. Here our academy has an advantage that no other one can compete with. I do not speak of the magnificence with which the king recompenses your labour, nor of the superb palace that he intends for you; he employs more reliable means to secure the glory of his Academy. These works that we have so often admired on days such as today will be /117/ eternal monuments of the esteem that he has for it, and of the case that he has made for its activities.

These, Gentlemen, are the advantages that each academician draws from the body of which he is a part; these are the motives that ought to excite him in a career in the sciences; and how powerfully so many motives combined together must act on you! Are those duties that the Academy imposes upon you anything other than what the love of the sciences alone would drive you to do? Could it be that you find too many constraints in the freest academy of Europe?

All the phenomena of nature, all the mathematical sciences, all the genres of literature are subject to your research, and as a result this company embraces a field far vaster than most other academies. But there are certain sanctuaries that no one is permitted to penetrate. Your founder himself, as sublime, as profound and as practised as he was in these ways, did not dare to lead his first disciples there. The legislators of all the academies, while ceding to them the entire nature of bodies, have forbidden that of spirits and speculation about first causes. Yet a monarch who has deigned to dictate our laws, a mind more vast, maybe more sure of your prudence, did not want to prohibit you from anything.

As for our academic discipline, there is no other academy in Europe whose rules require so little. For it would not be just to include in this comparison societies in which neither the eye nor the benefaction of the sovereign ever have any influence.

Our Academy embraces all of the sciences in four departments. Each class contributes equally to the progress of the others, however, the diversity of their objects admits a diversity in their manner of treatment. /118/

The first of our classes, that of experimental philosophy, encompasses all natural history, all knowledge for which we have need of eyes, of hands and of all the senses. It considers the bodies of the universe endowed with their sensible properties; it compares these properties, it binds them together and deduces them from each other. This science is completely founded upon experimentation. Without it, reason, always prone to being carried off to falsehoods, loses itself in systems that are refuted by it. However, experimentation also needs reason: it saves the natural philosopher time and pains; it makes him grasp all of a sudden certain relations that allow him to dispense with several useless activities, and permits him to turn all of his attention towards those phenomena that are decisive.

The natural philosopher should therefore apply himself to carefully examine experiments conducted by others; he should not be more indulgent with regard to his own; he should draw only legitimate consequences from them; and above all, equally removed from the ostentation which produces marvels and from the mystery that keeps the useful hidden, he should reveal them to his colleagues with all of their circumstances.

We see more than one academician whom I could cite here by way of a model: they know equally well the art of conducting the most delicate experiments and that of deriving the most ingenious consequences; in spite of being occupied with the noblest activities and the most useful matters of court and of the city, they find the time to give us excellent works and are the foremost and most assiduous in our meetings.

Our second class is the class of mathematics. The first considered bodies endowed with all their sensible properties. This one strips them of most of these properties so as to make a stricter and surer examination of those that remain. Bodies thus stripped present to the geometer only extension and number /119/ and those that immense distances place beyond the reach of several of his senses appear only more subject to his speculations and calculations.

Geometry, which owes its origin to its utility, and which the first geometers applied with so much success to the needs of life, was then for several centuries only sterile speculation and a kind of mind game. Too limited by its abstractions, it contented itself with exercising its art on difficult trifles, and did not dare to extend it to the phenomena of nature. The happy revolution brought about recently in our day in the sciences has rendered it more audacious. We have seen geometry explain all the phenomena of motion, and which part of this does not pertain to natural philosophy? We have seen it follow the ray of light in the space of the heavens, through all the bodies that it penetrates, calculate all the marvels which arise from its reflections and refractions; either to lead us to discover objects which because of their immense distance evaded our eyes, or to reveal to us those which by their extreme minuteness could not be perceived. We have seen the geometer determine by exact dimensions the size and figure of the globe that we inhabit, showing to the geographer the true position of all the places of the earth, instructing the navigator on the sure rules for reaching them. We have seen the mathematical sciences applied to all the useful or agreeable arts.

The march of the geometer is so determined, his steps are, so to speak, so counted, that there remain only a few pieces of advice to be given him.

The first is in the choice of subjects to which he applies himself, to keep more in view the utility of problems than their difficulty. How many geometers, if it is permissible to call them by this name, have /120/ lost time in the study of the quadrature of a curve which will never be traced!

The second piece of advice is that, in physico-mathematical problems that the geometer solves, he should always review the abstractions he has made; that his solutions are only right in so far as the bodies consist of only that small number of properties that he considers within them, and that, as in nature there are perhaps no bodies that are reduced solely to those properties, he ought, for those which have been the objects of his calculations, to consult the experiment again in order to discover whether those properties of which he has made an abstraction, or whose presence he has ignored, do not alter the effects of those that he has retained.

In following this advice, the geometer will place his art beyond the reproach of being useless, and he will justify it in the eyes of those who, not understanding it well, impute to him the faults that ought only be attributed to the mischievous use of it.

The class of speculative philosophy is the third. Experimental philosophy examined bodies as they are: endowed with all their sensible properties. Mathematics stripped them for the most part of these properties. Speculative philosophy considers those objects which no longer have any bodily property.

The Supreme Being, the human mind and all that which belongs to the mind are the objects of this science. The nature of bodies themselves, as represented by our perceptions, if they are different from these perceptions, is its concern.

But a fatal remark may be made, and one that we cannot avoid making: the more objects are interesting to us, the more difficult and uncertain is the knowledge that we can acquire about them! We will expose ourselves to many errors, and to very dangerous errors, if we do not use the greatest circumspection in this science which deals with minds. Let us guard ourselves from /121/ believing that by employing here the same method or the same words as in the mathematical sciences, we might arrive at the same certainty. This certainty is only attached to the simplicity of those objects that the geometer considers, those objects which admit only that which he has wished to suppose.

If I exhibited here the severity of the danger of those speculations that concern the Supreme Being, first causes and the nature of minds, it is not, Gentlemen, because I wish to divert you from these researches. All is permitted to the philosopher, provided that he treats all with the *esprit philosophique*, that is to say, with that mind which measures the different degrees of assent, which distinguishes evidence, probability, doubt, and which only presents its speculations under that of these different aspects to which they belong.

If the majority of objects that speculative philosophy considers appear so far above the forces of our minds, certain parts of this science are more within our reach. I speak of those duties that bind us to the Supreme Being, to other men, and to ourselves; of those laws to which all intelligent beings must be subject: a vast field and the most useful to cultivate! To them apply your care and watchfulness, but never forget, when you lack evidence, that another equally sure light must still guide you.

The fourth of our classes unites all the different objects of two celebrated academies of a kingdom where the abundance of great men has so multiplied them. I speak of our class of belles-lettres, which encompasses languages, history and all genres of literature: from the first elements of this art, which teaches the formation of sounds and signs to express thoughts, up to the extensive use that can be made of them. /122/

This art, the most marvellous of all, the most useful without doubt, was at its outset without doubt a very simple art. The few needs that the first men felt did not require a great number of words, nor of signs to express them. It was only after the success of this first attempt that they desired to communicate to each other some less common ideas, and began to understand the charms of conversation. How long it took, how many centuries passed before they knew how to paint for the eyes the conversation itself!

The first language of men was most probably already diversified when they passed from speaking to writing. Families became nations, and each by different sequences of ideas formed itself not only different words, but different manners of expression. Languages came from this diversity and all those children of the same father so dispersed and after so many generations, could no longer either recognise or understand each other when they met together again.

A fine project it would be, not to return them to their paternal language, a thing that is not possible, but to create for them a language more regular than all our languages which have formed only little by little, an easier language, and one that can be understood by all.

This problem which has been proposed more than once, was the object of our Academy from its birth: a clever man undertook the work:* a more able man regarded it as possible and did not undertake it.** This is not the place to set out the thoughts which have /123/ come to me on the subject; these would seem to belong more to one of the other classes than to this one.[2]

The most celebrated academy in the universe has, for a century, been occupied with perfecting that language which was already closest to perfection. The greatest geniuses of Antiquity and several of

* Solbrig [David Solbrig].

** Leibniz [G. W. Leibniz].

2. The clause 'these would seem [...] this one' is omitted in the 1752, 1755 and 1768 editions, as are the subsequent four paragraphs which are replaced by the following (pagination is from the 1768 edition):

> /298/ The multiplicity of the objects of this class [of speculative philosophy] does not permit me to give for each [of these objects] rules or advice. I will confine myself to making known the reason for the choice of the language in which our works appear, if it is still necessary to prove that the one ordained by he who is the soul of our Academy is the most suitable.
>
> The utility of academies is not contained within the limits of each nation. An academy possesses those men destined to enlighten the entire world: all the nations must benefit from their discoveries, and it is necessary that these be communicated to them in the most universal language. And no one, I believe, will deny this advantage to ours, which today seems to be the language of the whole of Europe, rather than the language of the French.
>
> If some other [language] could dispute the universal nature of French, it would be Latin. This language, it is true, is widespread everywhere: but it is dead, and is everywhere confined to a small number of scholars. We /299/ are only sure that we speak it well when we employ whole sentences by ancient authors; and as soon as we deviate from them we form a heterogeneous jargon whose ridiculousness we fail to perceive due to ignorance alone.
>
> The choice of our language can also be justified by other reasons that are no less strong; the very perfection of the language itself, the richness that our progress in all the arts and in all the sciences has introduced to it, the ease with which one can express oneself in it with accuracy on all sorts of subjects, the innumerable number of excellent books written in this language. If the Greeks and the Latins gave us the first models, these models have been surpassed in several genres; and in all genres multiplied so many times over that our works can today serve as models for writers from all nations.
>
> If one could make a criticism of our language, it is perhaps that which one made of the language of the Romans, when /300/ after attaining its greatest perfection, it came to lose its noble simplicity for that vain subtlety which we so inappropriately call *bel esprit*.

the moderns have given us certain rules for all the genres of writing. The length of this discourse does not permit me to deal with such subjects, even were I capable of it, so I limit myself to a few general principles.

The writer, in whatever genre of literature he applies himself, must never forget that words being the signs of ideas, the first point is the choice of the proper word. He should never permit himself to be seduced by harmony or measure, never should pleasantness or uneasiness make him say anything other than that which he wants to say.

The construction of a phrase forms a part[3] of the meaning of the words that are employed there. The writer, therefore, should religiously observe the rules of syntax.

Simple and pure style should also keep a distance from *pedantic gravity* and from that which we so inappropriately call *bel esprit.*

Some people[4] still cannot forgive a French author for denying the Germans the *bel esprit*;*[5] if they knew better what is usually meant by *bel esprit*, they would see that they have little reason to complain. It is most often merely the art of giving a common thought an affected turn of phrase. It is, so said one of the greatest Englishmen, *the art of making things appear more ingenious than they are.***[6] /124/

Some German authors have avenged themselves by denying our Frenchmen profundity and erudition. The vengeance would have been more just and no less irksome if in leaving us the *bel esprit*, they had contented themselves with saying that we make too much fuss over it. But if these authors understand by the erudition that they deny the French a jumble of Latin, Greek and Hebrew citations, a prolix and cluttered style, we will be grateful for their reproach and will applaud the defect.

This clarity of style which characterises our authors doubtless depends greatly upon the genius of our language, and it is this that

3. Reading 'partie' for 'parie'.
4. The 1768 edition has 'Quelques Savans des Universités'; see *Œuvres de Maupertuis*, vol.3, p.300.
* Father Bouhours [Dominique Bouhours].
5. The 1752 and 1755 editions omit this footnote.
** Bacon, *De augmentis scientiarum*, Book I [see Francis Bacon, *De dignitate et augmentis scientiarum*, in *Works of Francis Bacon*, ed. J. Spedding, R. L. Ellis and D. D. Heath, 7 vols (London, 1861-1879), vol.1, p.452].
6. The 1752, 1755 and 1768 editions do not provide the reference in the footnote to Bacon's *De augmentis scientiarum*.

has made it somehow the universal language of Europe. This is what makes a monarch whose taste is the most decisive vote, speak it and write it with such elegance, and to desire it to be the language of his Academy.

I have covered here all the different sciences to which we apply ourselves and I have not spoken of one which was one of the principal objects of this company when it was established.

The first rule of the Royal Society held that one of its classes had to apply itself to *the study of religion and the conversion of the infidels*. An article more singular for the manner in which it was presented than it may actually be. Our modern rule does not charge any class in particular with this activity, but who can deny that all of them contribute to it?

Can we not find in the study of the marvels of nature some proofs of the existence of a supreme being?

What is more capable of making us understand his wisdom than geometrical truths, than the eternal laws by which he rules the universe? /125/

Does not speculative philosophy make us see the necessity of the existence of an infinitely perfect being?[7]

Finally, the study of facts teaches us that he manifested himself to all human beings in an even more discernible manner, that he has demanded worship from them and has prescribed it to them.[8]

Translated by Peter R. Anstey and Fui Lee Luk

7. Where the 1750 edition has 'la nécessité de l'Existence d'un Etre infiniment parfait' the 1752, 1755 and 1768 editions have 'la nécessité de son Existence'.
8. Reading 'prescrit.' for 'prescrit?' as in the 1752 edition (p.147), and subsequent editions.

List of contributors

Peter R. Anstey is Professor of Philosophy at the University of Sydney. He specialises in early modern philosophy with a focus on individual philosophers such as Locke, Boyle and D'Alembert, as well as the broader themes of experimental philosophy and the theory of principles. He is the author of *John Locke and natural philosophy* (Oxford, 2011) and the editor of *The Oxford handbook of British philosophy in the seventeenth century* (Oxford, 2013).

Paola Basso is an independent researcher in philosophy (in Milan) and specialises in seventeenth- and eighteenth-century philosophy. Her research focuses on the relationship between metaphysics and geometry and on the philosophy of Johann Heinrich Lambert. Her publications include the monograph *Il secolo geometrico: la questione del metodo geometrico da Spinoza a Kant* (Florence, 2004).

Angela Ferraro is Assistant Professor at Université Laval (Quebec, Canada). She works mainly on early modern philosophy, especially Malebranche and the Enlightenment. She is coeditor of *Mémoires philosophiques de l'Académie de Berlin 1745-1769* (forthcoming) and has published *La Réception de la philosophie de Malebranche en France au XVIIIe siècle: métaphysique et épistémologie* (Paris, 2019).

Annelie Grosse is a post-doctoral researcher at the Berlin-Brandenburg Academy of Sciences and Humanities. She has worked and published on religious and moral philosophy at the Berlin Academy during the eighteenth century with a special focus on the work of Jean Henri Samuel Formey.

Christian Leduc is Professor of Philosophy at the Université de Montréal. He specialises in Leibniz and the German and French Enlightenment, and has recently coedited *Leibniz and natural teleology in the 18th century* (*Studia Leibnitiana* 50:1, 2018).

Ansgar Lyssy is working on a project on causality in Hegel at the University of Heidelberg. He is the author of *Kausalität und Teleologie bei G. W. Leibniz* (*Studia Leibnitiana* 48, 2016) and coeditor of *Kant on morality, humanity, and legality: practical dimensions of normativity* (Cham, 2021).

Alessandro Nannini is research fellow at the University of Bucharest. His current research is on the founding of aesthetics in the German Enlightenment, with a special focus on the philosophy of Baumgarten, Sulzer and Garve. He is the coeditor of the Italian edition of Baumgarten's *Aesthetics* (Milan, 2020).

Tinca Prunea-Bretonnet is Researcher in Philosophy at the Research Institute of the University of Bucharest and at the Institute for Philosophy and Psychology of the Romanian Academy. Her work focuses primarily on Kant and eighteenth-century German philosophy, as well as on the French Enlightenment and the reception of Kant's philosophy in the twentieth century. She is coeditor of *Kant et Wolff: héritages et ruptures* (Paris, 2011) and *The Experiential turn in eighteenth-century German philosophy* (New York, 2021).

Bibliography

Archival material

Archiv der Berlin-Brandenburgischen Akademie der Wissenschaften (BBAW), Berlin:
I. Historische Abteilung, Abt. I Nr. 5: Acta betr. die Verbindung der alten Societät mit der neuen Akademie der Wissenschaften 1743-1746.
I. Historische Abteilung, Abt. IV Nr. 12: Protocolle. Errichtung der neuen Akademie der Wissenschaften 1744-1746.

Archives de l'Académie des sciences et de l'Institut de France, Paris, 'Liste de l'Académie royale des sciences et des belles lettres de Berlin pour l'année MDCCXLVII', fonds 43 J, Pierre Louis Moreau de Maupertuis, dossier 63.

Basel Universitätsbibliothek, '5 lettres de Lambert et Sulzer', Handschriftlicher Nachlass, L.Ia 745, Johann H. Lambert, p.183-204.

Published sources

Primary sources

Arcy, Patrick d', 'Réflexions sur le principe de moindre action de M. de Maupertuis', *Mémoires de l'Académie des sciences de Paris* (1749/1753), p.531-38.

–, 'Réplique à un mémoire de M. de Maupertuis', *Mémoires de l'Académie des sciences de Paris* (1752/1756), p.503-19.

Aristotle, *The Complete works of Aristotle*, ed. Jonathan Barnes, 2 vols (Princeton, NJ, 1984).

Bacon, Francis, *The Advancement of learning*, ed. M. Kiernan, in *The Oxford Francis Bacon*, vol.4 (Oxford, 2000).

–, *Works of Francis Bacon*, ed. J. Spedding, R. L. Ellis and D. D. Heath, 7 vols (London, 1861-1879).

Baumgarten, Alexander G., *Metaphysics* (1st edn 1739; 4th edn 1757), ed. and translated by Courtney D. Fugate and John Hymers (London, 2013).

Beausobre, Louis de, *Dissertations philosophiques dont la première roule sur la nature du feu et la seconde sur les différentes parties de la philosophie, et des mathématiques* (Paris, Durand and Pissot, 1753).

–, *Le Pyrrhonisme raisonnable* (Berlin, Etienne de Bordeaux, 1755).

–, 'Réflexions philosophiques sur les songes, premier mémoire', in *Histoire de l'Académie royale des sciences et belles lettres* (1762; Berlin, Haude and Spener, 1769), p.429-40.

–, 'Réflexions sur la nature et la nécessité des idées obscures', in *Histoire de l'Académie royale des sciences et belles lettres* (1768; Berlin, Haude and Spener, 1770), p.411-18.

–, 'Réflexions sur la nature et les causes de la folie, premier mémoire', in *Histoire de l'Académie royale des sciences et belles lettres* (1759; Berlin, Haude and Spener, 1766), p.390-403.

Béguelin, Nicolas, 'Les premiers principes de la métaphysique', in *Histoire de l'Académie royale des sciences et belles lettres* (1755; Berlin, Haude and Spener, 1757), p.405-23.

Boerhaave, Herman, *Elementa chemiæ*, vol.1 (Leiden, Isaacum Severinum, 1732).

Boyle, Robert, *Usefulness of experimental natural philosophy* (Oxford, Richard Davis, 1663).

–, *The Works of Robert Boyle*, ed. Michael Hunter and Edward B. Davis, 14 vols (London, 1999-2000).

A Brief vindication of the Royal Society (London, John Martin, 1670).

Brucker, Johann Jakob, *Historia critica philosophiae*, vol.4.2 (Leipzig, Breitkopf, 1744).

Budde(us), Johann Franz, *Elementa philosophiae instrumentalis seu Institutiones philosophiae eclecticae*, 10th edn, vol.1 (Halle, Orphanotropheum, 1755).

Buffon, Georges-Louis L., comte de, 'Initial discourse', in *From natural history to the history of nature*, translated by John Lyon and Phillip Sloan (Notre Dame, IN, 1980), p.100-103.

–, 'Preface' to *La Statique des vegetaux, et l'analyse de l'air* by Stephen Hales, in *From natural history to the history of nature*, translated by John Lyon and Phillip Sloan (Notre Dame, IN, 1980), p.35-40.

Castillon, Jean de, 'Essai d'une théorie métaphysico-mathématique de l'expérience', in *Mémoires de l'Académie royale des sciences et belles lettres* (1796; Berlin, Decker, 1797), p.364-90.

Chambers, Ephraim, *Cyclopaedia* (London, James and John Knapton, 1728).

Cicero, *Tusculan disputations*, translated by J. E. King (Cambridge, MA, 1945).

Clarke, Samuel, *A Collection of papers, which passed between the late learned Mr. Leibniz, and Dr. Clarke, in the years 1715 and 1716* (London, James Knapton, 1717).

Condillac, Etienne Bonnot de, *Essai sur l'origine des connaissances humaines*, in *Œuvres philosophiques de Condillac* (1746), ed. Georges Leroy, vol.1 (Paris, 1947).

–, *Essay on the origin of human knowledge*, translated by Hans Aarsleff (Cambridge, 2001).

–, *Les Monades*, ed. Laurence L. Bongie, *SVEC* 187 (1980).

–, *Œuvres philosophiques*, ed. Georges Le Roy, vol.1 (Paris, 1947).

–, *Philosophical writings of Etienne Bonnot de Condillac*, translated by Franklin Philip (Hillsdale, NJ, 1982).

–, *Traité des systèmes* (The Hague, Neaulme, 1749).

Cotes, Roger, *Leçons de physique expérimentale*, translated by Pierre Le Monnier (Paris, David, 1742).

D'Alembert, Jean, 'Causes finales (*metaph.*)', in *Encyclopédie, ou Dictionnaire raisonné des sciences, des arts et des métiers*, ed. Denis Diderot and Jean D'Alembert, vol.2 (Paris, Briasson *et al.*, 1751), p.789.

–, 'Discours préliminaire' (1751), in *Encyclopédie, ou Dictionnaire raisonné des sciences, des arts et des métiers*, ed. Denis Diderot and Jean D'Alembert, University of Chicago: ARTFL Encyclopédie Project (autumn 2017 edn), ed. Robert Morrissey and Glenn Roe, http://encyclopedie.uchicago.edu/ (last accessed 13 April 2022).

–, *Essai sur les éléments de philosophie* (Hildesheim, 2003).

–, *Essai sur les eléments de philosophie*, in *Mélanges de littérature, d'histoire et de philosophie*, vol.4 (Amsterdam, Zacharis Chatelain et fils, 1759).

–, *Preliminary discourse to the Encyclopedia of Diderot*, translated by Richard N. Schwab (Chicago, IL, 1995).

–, *Traité de dynamique* (Paris, David, 1743).

–, *Traité de dynamique*, 2nd edn (Paris, chez David l'aîné, 1758).

Desaguliers, John Theophilus, *Lectures of experimental philosophy* (London, W. Mears, 1719).

–, *Physico-mechanical lectures, or an Account of what is explain'd and demonstrated in the course of mechanical and experimental philosophy given by J. T. Desaguliers* (London, Richard Bridger, 1717).

Descartes, René, *The Philosophical writings of Descartes*, vol.1, ed. John Cottingham, Robert Stoothoff and Dugald Murdoch (Cambridge, 1985).

Desfontaines, Pierre-François Guyot, and François Granet (ed.), *Observations sur les écrits modernes, tome dix-septième* (Paris, Chaubert, 1739).

Des Maizeaux, Pierre (ed.), Recueil de pièces diverses sur la philosophie, la religion et l'histoire par Leibniz, Clarke et Newton (Amsterdam, H. Du Sauzet, 1720).

Dictionnaire philosophique, ou Introduction à la connoissance de l'homme (London [i.e. Paris], n.n., 1751).

Diderot, Denis, 'Eclectisme', in *Encyclopédie, ou Dictionnaire raisonné des sciences, des arts et des métiers*, ed. Denis Diderot and Jean D'Alembert, vol.5 (Paris, Briasson *et al.*, 1755), p.270-93.

–, *Œuvres complètes de Diderot*, ed. Jules Assézat, 7 vols (Paris, 1875-1877).

–, and Jean D'Alembert (ed.), *Encyclopédie, ou Dictionnaire raisonné des sciences, des arts et des métiers*, University of Chicago: ARTFL Encyclopédie Project (autumn 2017 edn), ed. Robert Morrissey and Glenn Roe, http://encyclopedie.uchicago.edu/ (last accessed 13 April 2022).

Dubos, Jean-Baptiste, *Réflexions critiques sur la poésie et sur la peinture* (1719; Paris, 1993).

Du Châtelet, Emilie, *Dissertation sur la nature et la propagation du feu* (Paris, Prault fils, 1744).

–, *Institutions de physique* (Paris, Prault fils, 1740).

Dunton, John, *The Young-students-library* (London, John Dunton, 1692).

Eller, Johann T., 'Seconde dissertation sur les élémens', in *Histoire de l'Académie royale des sciences et belles lettres* (1746; Berlin, Haude, 1748), p.25-48.

Euler, Leonhard, *Anleitung zur Naturlehre*, in *Opera omnia*, series 3, vol.1 (Leipzig, 1926), p.16-180.

–, *Anleitung zur Naturlehre*, in *Opera postuma mathematica et physica*, ed. Paulus Henricus Fuss and Nicolaus Fuss, 2 vols (St Petersburg, 1862), vol.2, p.449-560.

–, *Commercium epistolicum*, vol.6: *Correspondance de Leonard Euler avec de Maupertuis et Frédéric II*, in *Opera omnia*, ed. Pierre Costabel, Eduard Winter, Ašot T. Grigorijan and Adolf P. Juškevič, series 4.a (Basel, 1986).

–, 'Découverte d'un nouveau principe de mécanique', in *Histoire de l'Académie royale des sciences et belles lettres* (1750; Berlin, Haude and Spener, 1752), p.185-217.

–, 'Essai d'une démonstration métaphysique du principe général de l'équilibre', in *Histoire de l'Académie royale des sciences et belles lettres* (1751; Berlin, Haude and Spener, 1753), p.246-54.

–, *Gedancken von den Elementen der Cörper* (Berlin, Haude and Spener, 1746); reprinted in *Opera omnia*, series 3, vol.2 (Basel, 1942), p.347-66.

–, *Letters of Euler on different subjects in physics and philosophy addressed to a German princess*, translated by Henry Hunter, 2nd edn, 2 vols (London, 1802).

–, *Lettres à une princesse d'Allemagne sur divers sujets de physique et de philosophie*, vol.2 (St Petersburg, Académie impériale des sciences, 1768); 2nd edn, 2 vols (Leipzig, Steidel, 1770); reprinted in *Opera omnia*, series 3, vol.11-12 (Basel, 1960).

–, *Mechanica sive motus scientia analytice exposita* (St Petersburg, Petropoli, 1736).

–, *Methodus inveniendi lineas curvas maximi minimive proprietate gaudentes, additamentum* 1: *De curvis elasticis* (Lausanne and Geneva, Bousquet, 1744), p.245-310; reprinted in *Opera omnia*, ed. A. Speiser, series 1, vol.24 (Bern, 1952).

–, *Methodus inveniendi lineas curvas maximi minimive proprietate gaudentes, additamentum* 2:

De motu projectorum in medio non resistente (Lausanne and Geneva, Bousquet, 1744), p.311-20.
–, *Opera omnia*, series 4.a, vol.6 (Basel, 1986).
–, 'Recherches sur les plus grands et plus petits qui se trouvent dans les actions des forces', in *Histoire de l'Académie royale des sciences et belles lettres* (1750; Berlin, Haude and Spener, 1752), p.149-88.
–, 'Recherches sur l'origine des forces', in *Histoire de l'Académie royale des sciences et belles lettres* (1750; Berlin, Haude and Spener, 1752), p.419-47.
–, 'Réflexions sur l'espace et le tems', in *Histoire de l'Académie royale des sciences et belles lettres* (1748; Berlin, Haude and Spener, 1750), p.324-33; reprinted in *Opera omnia*, series 3, vol.2 (Basel, 1942), p.376-83.
–, 'Sur le principe de la moindre action', in *Histoire de l'Académie royale des sciences et belles lettres* (1751; Berlin, Haude and Spener, 1753), p.199-218.
–, *Theoria motus corporum solidorum seu rigidorum* (1765), in *Opera omnia*, series 2, vol.3 (Basel, 1948).
–, and Johann H. Lambert, *L. Euler und J. H. Lambert Briefwechsel*, ed. Karl Bopp (Berlin, 1924).

Fontenelle, Bernard Le Bovier de, *Histoire du renouvellement de l'Académie royale des sciences en MDCXCIX* (Paris, Jean Boudot, 1708).
Formey, Jean Henri Samuel, *La Belle Wolfienne*, 6 vols (The Hague, Le Vier, 1741, and Neaulme, 1743-1753).
–, 'Considérations sur ce qu'on peut regarder aujourd'hui comme le but principal des académies, et comme leur effet le plus avantageux', in *Histoire de l'Académie royale des sciences et belles lettres* (1767; Berlin, Haude and Spener, 1769), p.367-81.
–, 'Considérations sur ce qu'on peut regarder aujourd'hui comme le but principal des académies, et comme leur effet le plus avantageux: second discours', in *Histoire de l'Académie royale des sciences et belles lettres* (1768; Berlin, Haude and Spener, 1770), p.357-66.
–, 'Discours sur ces questions: Quel est le degré de certitude dont sont susceptibles les preuves tirées de la considération de cet univers pour demontrer l'existence d'une divinité? Et quelle est la meilleur maniere de faire usage de ces argumens a posteriori, pour établir cette importante vérité?', *Nouvelle bibliothèque germanique* 25 (1759), p.317-41; reprinted in *Histoire de l'Académie royale des sciences et belles lettres* (1765; Berlin, Haude and Spener, 1767), p.435-49.
–, *Elementa philosophiae seu medulla wolfiana in usum auditorum* (Berlin, Haude and Spener, 1746).
–, 'Examen de la preuve qu'on tire des fins de la nature, pour établir l'existence de Dieu', in *Histoire de l'Académie royale des sciences et belles lettres* (1747; Berlin, Haude and Spener, 1749), p.365-84.

–, 'Examen philosophique de la liaison réelle qu'il y a entre les sciences et les moeurs', in *Histoire de l'Académie royale des sciences et belles lettres* (1753; Berlin, Haude and Spener, 1755), p.397-416.

–, 'Introduction à la métaphysique, ou dissertation préliminaire sur les notions directrices & sur le véritable usage de l'ontologie', in *La Belle Wolfienne*, vol.4 (The Hague, Neaulme, 1746), p.i-xxiii.

[–], 'Préface', in *Histoire de l'Académie royale des sciences et des belles lettres de Berlin* (1745; Berlin, Haude, 1746), sig. [)(4]-)()(2*v*.

–, 'Les preuves de l'existence de Dieu, ramenées aux notions communes', in *Histoire de l'Académie royale des sciences et belles lettres* (1747; Berlin, Haude and Spener, 1749), p.341-64.

Frederick II, king of Prussia, 'De la littérature allemande', in *Œuvres philosophiques* (Paris, 1985), p.423-55.

–, *Letters between Frederick II and M. D'Alembert*, in *Posthumous works of Frederick II*, translated by Thomas Holcroft, vol.11 (London, G. G. J. and J. Robinson, 1789).

Girard, Gabriel, *Synonymes français: leurs différentes significations, et le choix qu'il en faut faire pour parler avec justesse* (Paris, Imprimerie de la veuve d'Houry, 1736).

Heinius, Johann Philipp, 'Vie d'Anaxagore, philosophe très célèbre dans l'école ionique', in *Histoire de l'Académie royale des sciences et belles lettres* (1752; Berlin, Haude and Spener, 1754), p.313-49.

Herschel, William, 'On the direction and velocity of the motion of the sun, and solar system', *Philosophical transactions* 95 (1805), p.233-56.

Huygens, Christiaan, *Œuvres complètes de Christiaan Huygens*, 22 vols (The Hague, 1888-1950).

Jariges, Philipp Joseph von, 'Examen du spinozisme et des objections de Mr. Bayle contre ce système', in *Histoire de l'Académie royale des sciences et des belles lettres de Berlin* (1745; Berlin, Haude, 1746), p.121-42.

Kahle, Ludwig M., *Vergleichung der Leibnitzischen und Neutonischen Metaphysik* (Göttingen, Königl. Universitäts-Buchhandlung, 1740).

Kant, Immanuel, 'Inquiry concerning the distinctness of the principles of natural theology and morality', in *Theoretical philosophy, 1755-1770*, ed. and translated by David Walford and Ralf Meerbrote (Cambridge, 1992), p.245-75.

–, 'What is orientation in thinking?', in *Kant's political writings*, ed. Hans Reiss (Cambridge, 1991), p.237-49.

Krüger, Johann G., *Grundriß eines neuen Lehrgebäudes der Artzneygelahrheit* (Halle, Hemmerde, 1745).

–, *Naturlehre*, 2nd edn, 3 vols (Halle, Hemmerde, 1744-1755).

Lambert, Johann H., *Anlage zur Architectonic* (Riga, Hartknoch, 1771); reprinted in *Philosophische Schriften*, vol.3 (Hildesheim, 1965).

–, *Briefwechsel*, in *Philosophische Schriften*, vol.9 (Hildesheim, 1968).

–, *Cosmological letters on the arrangement of the world-edifice*, translated by Stanley L. Jaki (New York, 1976).

–, 'Discours de réception', in *Histoire de l'Académie royale des sciences et belles lettres* (1765; Berlin, Haude and Spener, 1767), p.506-14.

–, *Neues Organon oder Gedanken über die Erforschung und Bezeichnung des Wahren und dessen Unterscheidung vom Irrtum und Schein* (Leipzig, Johann Wendler, 1764); reprinted in *Philosophische Schriften*, vol.1 (Hildesheim, 1965).

–, 'Observations sur quelques dimensions du monde intellectuel', in *Histoire de l'Académie royale des sciences et belles lettres* (1763; Berlin, Haude and Spener, 1770), p.421-43.

–, *Theorie der Zuverläßigkeit der Beobachtungen und Versuche*, in *Beyträge zum Gebrauche der Mathematik und deren Anwendung* (Berlin, Buchladen der Realschule, 1792).

–, 'Vom Trabanten der Venus', in *Astronomisches Jahrbuch* (Berlin, 1777), p.186-91.

La Mettrie, Julien Offray de, *Traité de l'âme*, ed. Theo Verbeek, vol.1 (Utrecht, 1988).

Leibniz, Gottfried Wilhelm, *Discourse on metaphysics*, in *Philosophische Schriften*, ed. C. I. Gerhardt, vol.4 (Hildesheim, 1978), p.427-63; reprinted in *Sämtliche Schriften und Briefe*, ed. Berlin-Brandenburgische Akademie der Wissenschaften, vol.VI.4 (Berlin, 1999), p.1529-88.

–, *G. W. Leibniz: philosophical essays*, ed. and translated by Roger Ariew and Daniel Garber (Indianapolis, IN, 1989).

–, *Œuvres philosophiques de feu Mr de Leibniz*, ed. Rudolph Eric Raspe (Amsterdam and Leipzig, Schreuder, 1765).

–, *Die philosophischen Schriften*, ed. C. I. Gerhardt, 7 vols (Hildesheim, 1965).

–, *Sämtliche Schriften und Briefe*, ed. Deutsche Akademie der Wissenschaften (Darmstadt, 1923-).

–, 'Unicum opticae, catoptricae et dioptricae principium', *Acta eruditorum* (June 1682), p.185-90.

Lichtenberg, Georg C., *The Waste books*, translated by R. J. Hollingdale (New York, 1990).

Locke, John, *Essai philosophique concernant l'entendement humain*, translated by Pierre Coste (1700), ed. Philippe Hamou (Paris, 2009).

–, *An Essay concerning human understanding* (1690), ed. Peter H. Nidditch (Oxford, 1975).

Maupertuis, Pierre-Louis Moreau de, 'Accord de différentes loix de la nature qui avoient jusqu'ici paru incompatibles', in *Memoires de l'Académie royale des sciences* (Paris, De l'imprimerie royale, 1748), p.417-26; reprinted in

Œuvres de Maupertuis, 4 vols (Lyon, Bruyset, 1768), vol.4, p.1-28; loosely translated in *The Gentleman's magazine and historical chronicle* 21 (May 1751), p.218-22.

–, 'Des devoirs de l'académicien', in *Histoire de l'Académie royale des sciences et belles lettres, depuis son origine jusqu'à présent, avec les pieces originales* (Berlin, Haude and Spener, 1750), p.115-25; reprinted in *Œuvres de Maupertuis*, 4 vols (Lyon, Bruyset, 1768), vol.3, p.283-302.

–, 'Discours prononcé le jour de la naissance du roi (1747)', in *Histoire de l'Académie royale des sciences et belles lettres* (1746; Berlin, Haude, 1748), p.10-16.

–, *Eloge de Montesquieu*, in *Œuvres de Maupertuis*, 4 vols (Lyon, Bruyset, 1768), vol.3, p.259-433.

–, *Essai de cosmologie* (Berlin, n.n., 1750); reprinted in *Œuvres de Maupertuis*, 4 vols (Lyon, Bruyset, 1768), vol.1, p.1-78.

–, *Essai de philosophie morale*, in *Œuvres de Maupertuis*, 4 vols (Lyon, Bruyset, 1768), vol.1, p.171-252.

–, 'Examen philosophique de la preuve de l'existence de Dieu employée dans l'*Essai de cosmologie*', in *Histoire de l'Académie royale des sciences et belles lettres* (1756; Berlin, Haude and Spener, 1758), p.389-424.

–, 'Expériences sur les scorpions', in *Memoires de l'Académie royale des sciences* (Paris, De l'imprimerie royale, 1733), p.223-29.

–, *Lettre sur le progrès des sciences*, in *Œuvres de Maupertuis*, 4 vols (Lyon, Bruyset, 1768), vol.2, p.373-431.

–, 'Les loix du movement et du repos déduites d'un principe métaphysique', in *Histoire de l'Académie royale des sciences et belles lettres* (1746; Berlin, Haude, 1748), p.267-94.

–, *Œuvres de Maupertuis*, 4 vols (Lyon, Bruyset, 1768).

–, 'Recherche des lois du mouvement' (1746), in *Œuvres de Maupertuis*, 4 vols (Lyon, Bruyset, 1768), vol.4, p.29-42.

–, 'Réponse à un mémoire de M. d'Arcy inséré dans le volume de l'Académie royale de Paris pour l'année 1749', in *Histoire de l'Académie royale des sciences et belles lettres* (1752; Berlin, Haude and Spener, 1754), p.293-98.

Meckel, Johann Friedrich, 'Recherches anatomico-physiologiques sur les causes de la folie qui viennent du vice des parties internes du corps humain', in *Histoire de l'Académie royale des sciences et belles lettres* (1764; Berlin, Haude and Spener, 1766), p.65-88.

Mercure de France (October 1738).

Merian, Jean Bernard, 'Abridgement of Moses Mendelssohn's prize-winning essay on the distinctness of the principles of metaphysics and natural theology read at the public meeting of the Prussian Royal Academy of Sciences on 2 June 1763 by M. Merian', in Immanuel Kant, *Theoretical philosophy, 1755-1770*, ed. and translated by David

Walford and Ralf Meerbrote (Cambridge, 1992), p.276-86.
–, *Discours sur la métaphysique* (Berlin, Voss, 1765); reprinted in *Histoire de l'Académie royale des sciences et belles lettres* (1765; Berlin, Haude and Spener, 1767), p.450-74.
–, 'Mémoire sur l'apperception considérée relativement aux idées', in *Histoire de l'Académie royale des sciences et belles lettres* (1749; Berlin, Haude and Spener, 1751), p.442-77.
–, 'Mémoire sur l'apperception de sa propre existence', in *Histoire de l'Académie royale des sciences et belles lettres* (1749; Berlin, Haude and Spener, 1751), p.416-41.
–, 'Parallèle historique de nos deux philosophies nationales', in *Mémoires de l'Académie royale des sciences et belles lettres* (1797; Berlin, Decker, 1800), p.53-96.

Newton, Sir Isaac, 'An account of the book entituled *Commercium epistolicum collinii & aliorum*', *Philosophical transactions* 29 (1714-1716), p.173-224.
–, *The Principia: mathematical principles of natural philosophy*, translated by I. Bernard Cohen and Anne Whitman (Berkeley, CA, 1999).
Nollet, Jean-Antoine (abbé), *Cours de physique expérimentale chez M. l'abbé Nollet* (Paris, P. G. Le Mercier, 1735).
–, *Programme, ou Idée générale d'un cours de physique expérimentale* (Paris, P. G. Le Mercier, 1738).

Oldenburg, Henry, *Correspondence of Henry Oldenburg*, ed. A. Rupert Hall and Marie Boas Hall, 13 vols (Madison, WI, 1965-1986).

Power, Henry, *Experimental philosophy* (London, John Martin and James Allestry, 1664).
Prémontval, Pierre Le Guay de, 'De la psychocratie: fin' (part 3), in *Histoire de l'Académie royale des sciences et belles lettres* (1761; Berlin, Haude and Spener, 1768) p.371-420.

Sack, August F. W., 'Sur de nouvelles pétrifications marines', in *Histoire de l'Académie royale des sciences et des belles-lettres de Berlin* (1745; Berlin, Haude, 1746), p.67-73.
Sprat, Thomas, *The History of the Royal-Society of London* (London, J. Martyn, 1667).
Sulzer, Johann G., *Allgemeine Theorie der Schönen Künste* (1771-1774), 4 vols (Leipzig, Weidmanns Erben und Reich, 1778-1779).
–, 'Analyse de la raison', in *Histoire de l'Académie royale des sciences et belles lettres* (1758; Berlin, Haude and Spener, 1765), p.414-42.
–, 'Analyse du génie', in *Histoire de l'Académie royale des sciences et belles lettres* (1757; Berlin, Haude and Spener, 1759), p.392-404.
–, 'Considérations psychologiques sur l'homme moral', in *Histoire de l'Académie royale des sciences et belles lettres* (1769; Berlin, Voss, 1771), p.361-80.

–, 'De l'énergie dans les ouvrages des beaux-arts', in *Histoire de l'Académie royale des sciences et belles lettres* (1765; Berlin, Haude and Spener, 1767), p.475-92.

–, 'Explication d'un paradoxe psychologique: que non seulement l'homme agit et juge quelquefois sans motifs et sans raisons apparentes, mais même malgré des motifs pressans et des raisons convainquantes', in *Histoire de l'Académie royale des sciences et belles lettres* (1759; Berlin, Haude and Spener, 1766), p.433-50.

–, *Kurzer Begriff aller Wißenschaften und andern Theile der Gelehrsamkeit, worin jeder nach seinem Inhalt, Nutzen und Vollkommenheit kürzlich beschrieben wird* (1745), 2nd edn (Frankfurt and Leipzig, Langenheim, 1759).

–, 'Observations sur les divers états où l'âme se trouve en exerçant ses facultés primitives, celle d'apercevoir et celle de sentir', in *Histoire de l'Académie royale des sciences et belles lettres* (1763; Berlin, Haude and Spener, 1770), p.407-20.

–, 'Observations sur l'influence réciproque de la raison sur le langage et du langage sur la raison', in *Histoire de l'Académie royale des sciences et belles lettres* (1767; Berlin, Haude and Spener, 1769), p.413-38.

–, 'Observations sur quelques propriétés de l'âme comparées à celles de la matière: pour servir à l'examen du matérialisme', in *Nouveaux mémoires de l'Académie royale des sciences et belles-lettres de Berlin* (1771; Berlin, Voss, 1773), p.390-410.

–, *Pensées sur l'origine et les différens emplois des sciences et des beaux-arts* (Berlin, Haude and Spener, 1757).

–, 'Recherches sur l'origine des sentimens agréables et désagréables, première partie: théorie générale du plaisir', in *Histoire de l'Académie royale des sciences et belles lettres* (1751; Berlin, Haude and Spener, 1753), p.57-75.

–, 'Recherches sur l'origine des sentimens agréables et désagréables, seconde partie: théorie des plaisirs intellectuels', in *Histoire de l'Académie royale des sciences et belles lettres* (1751; Berlin, Haude and Spener, 1753), p.76-100.

–, 'Recherches sur l'origine des sentimens agréables et désagréables, troisième partie: des plaisirs des sens', in *Histoire de l'Académie royale des sciences et belles lettres* (1752; Berlin, Haude and Spener, 1754), p.350-72.

–, 'Recherches sur l'origine des sentimens agréables et désagréables, quatrième partie: des plaisirs moraux', in *Histoire de l'Académie royale des sciences et belles lettres* (1752; Berlin, Haude and Spener, 1754), p.373-90.

–, 'Recherches sur un principe fixe, qui serve à distinguer les devoirs de la morale de ceux du droit naturel', in *Histoire de l'Académie royale des sciences et belles lettres* (1756; Berlin, Haude and Spener, 1758), p.450-56.

–, 'Réflexions philosophiques sur l'utilité de la poésie dramatique', in *Histoire de l'Académie royale des sciences et belles lettres* (1760; Berlin, Haude and Spener, 1767), p.326-40.

–, 'Sur l'apperception et son influence sur nos jugemens', in *Histoire de l'Académie royale des sciences et belles lettres* (1764; Berlin, Haude and Spener, 1766), p.415-34.

–, 'Sur l'immortalité de l'âme considérée physiquement: premier et second mémoire', in *Nouveaux mémoires de l'Académie royale des sciences et belles-lettres* (1775; Berlin, Voss, 1777), p.359-87.

–, 'Sur l'immortalité de l'âme considérée physiquement: troisième mémoire', in *Nouveaux mémoires de l'Académie royale des sciences et belles-lettres* (1776; Berlin, Voss, 1779), p.349-59.

–, *Teoria generale delle Belle Arti* (1771-1774), partial Italian edition by Alessandro Nannini (Bologna, 2011).

Thomasius, Christian, *Einleitung zur Hof-Philosophie* (Frankfurt and Leipzig, Rüdger and Bauern, 1710).

Voltaire, *Lettres philosophiques* (Amsterdam, E. Lucas, 1734).

–, *Traité de métaphysique* (1734), in *Œuvres complètes de Voltaire*, vol.22 (Paris, 1879).

Walch, Johann G., *Philosophisches Lexikon* (Leipzig, Gleditsch, 1726).

Winter, Eduard, *Die Registres der Berliner Akademie der Wissenschaften 1746-1766: Dokumente für das Wirken Leonhard Eulers in Berlin zum 250. Geburtstag* (Berlin, 1957).

Wolff, Christian, 'De notionibus directricibus & genuino usu philosophiae primae', in *Horae subsecivae Marburgenses anni MDCCXXIX, quibus philosophia ad publicam privatamque utilitatem aptatur*, in *Gesammelte Werke*, 1st edn (1729), section II, vol.34.1 (Hildesheim, 1983), p.310-50.

–, *Discursus praeliminaris de philosophia in genere / Einleitende Abhandlung über Philosophie im Allgemeinen. Historisch-kritische Ausgabe*, ed. Günter Gawlick and Lothar Kreimendahl (Stuttgart and Bad Cannstatt, 1996).

–, *Philosophia prima, sive Ontologia* (Frankfurt and Leipzig, Renger, 1729); reprinted in *Gesammelte Werke*, section II, vol.3 (Hildesheim, 2011).

–, *Philosophia rationalis, sive Logica* (Frankfurt, Renger, 1728).

–, *Psychologia empirica* (Frankfurt, Renger, 1732); reprinted in *Gesammelte Werke*, section II, vol.5, ed. Jean Ecole *et al.* (Hildesheim, 1968).

–, *Psychologia rationalis* (Frankfurt and Leipzig, Renger, 1734).

–, *Vernünfftige Gedancken von der Kräfften des menschlichen Verstandes* (Halle, Renger, 1713).

–, 'Vorbericht von der Welt-Weisheit', in *Vernünfftige Gedancken von den Kräften des menschlischen Verstandes und ihrem richtigen Gebrauche in Erkäntniss der Wahrheit*, in *Gesammelte Werke*, 4th edn, section I, vol.1 (Hildesheim, 2006), p.115-22.

Secondary sources

Aarsleff, Hans, 'The Berlin Academy under Frederick the Great', *History of the human sciences* 2:2 (1989), p.193-206.

Ahnert, Thomas, 'Newtonianism in early Enlightenment Germany, c. 1720 to 1750: metaphysics and the critique of dogmatic philosophy', *Studies in history and philosophy of science* 35:3 (2004), p.471-91.

Albrecht, Michael, *Eklektik: eine Begriffsgeschichte mit Hinweisen auf die Philosophie- und Wissenschaftsgeschichte* (Stuttgart and Bad Cannstatt, 1994).

–, 'Thomasius – kein Eklektiker?', in *Christian Thomasius 1655-1728: Interpretationen zu Werk und Wirkung*, ed. Werner Schneiders (Hamburg, 1989), p.73-94.

Anstey, Peter R., 'Bacon, experimental philosophy and French Enlightenment natural history', in *Natural history in early modern France: the poetics of an epistemic genre*, ed. Raphaële Garrod and Paul J. Smith (Leiden, 2018), p.205-40.

–, 'D'Alembert, the "Preliminary discourse" and experimental philosophy', *Intellectual history review* 24:4 (2014), p.495-516.

–, 'Experimental pedagogy and the eclipse of Robert Boyle in England', *Intellectual history review* 25:1 (2015), p.115-31.

– (ed.), *The Idea of principles in early modern thought* (New York, 2017).

–, *John Locke and natural philosophy* (Oxford, 2011).

–, 'The principled Enlightenment: Condillac, D'Alembert and principle minimalism', in *Rethinking the Enlightenment: between history, philosophy, and politics*, ed. Geoff Boucher and Henry Martyn Lloyd (Lanham, MD, 2018), p.131-50.

–, and Dana Jalobeanu, 'Experimental natural history', in *The Cambridge history of the philosophy of the scientific revolution*, ed. David M. Miller and Dana Jalobeanu (Cambridge, 2022), p.222-37.

–, and Alberto Vanzo, 'The origins of early modern experimental philosophy', *Intellectual history review* 22:4 (2012), p.499-518.

Bærtschi, Bernard, 'La conception de la conscience développée par Mérian', in *Schweizer im Berlin des 18. Jahrhunderts*, ed. Martin Fontius and Helmut Holzhey (Berlin,1996), p.217-30.

–, *Conscience et réalité: études sur la philosophie française au XVIIIe siècle* (Geneva, 2005).

Baillot, Anne, 'Louis de Beausobre, entre cour et académie: la correspondance d'un intellectuel francophone en Prusse au XVIIIe siècle', in *Entrer en communication de l'âge classique aux Lumières*, ed. Pierre-Yves Beaurepaire and Héloïse Hermant (Paris, 2013), p.123-40.

–, 'Pyrrhonismus und Politik an der Akademie der Wissenschaften um 1750', in *Französisch-deutsche Kulturräume um 1800.*

Bildungsnetzwerke – Vermittlerpersönlichkeiten – Wissenstransfer, ed. Anna Busch, Nana Hengelhaupt and Alix Winter (Berlin, 2012), p.25-42.

Barber, W. H., *Leibniz in France: from Arnauld to Voltaire* (Oxford, 1955).

Barth, Christian, 'Leibnizian conscientia and its Cartesian roots', *Studia Leibnitiana* 43:2 (2011), p.216-36.

Bartholmèss, Christian, *Histoire philosophique de l'Académie de Prusse depuis Leibniz jusqu'à Schelling, particul. sous Frédéric-le-Grand*, 2 vols (Paris, 1850).

Basso, Paola, 'Deutlichkeit, évidence et certitude dans les réponses anonymes', *Astérion* 9 (2011), doi:10.4000/asterion.2101.

–, 'Lambert on experience and deduction', in *The Experimental turn in eighteenth-century German philosophy*, ed. Karin de Boer and Tinca Prunea-Bretonnet (New York, 2021), p.181-202.

–, 'Le mythe de la démonstrabilité résiste-t-il encore? Remarques sur l'orientation des réponses anonymes', *Astérion* 9 (2011), doi:10.4000/asterion.2112.

Batteux, Charles, *The Fine arts reduced to a single principle*, translated by James O. Young (Oxford, 2015).

Beck, Lewis W., *Early German philosophy: Kant and his predecessors* (Cambridge, 1969).

Beeson, David, *Maupertuis: an intellectual biography*, *SVEC* 299 (1992).

Blumenberg, Hans, *The Genesis of the Copernican world*, translated by R. Wallace (Cambridge, MA, 1987).

Bödeker, Hans-Erich, 'Von der "Magd der Theologie" zur "Leitwissenschaft": Vorüberlegungen zu einer Geschichte der Philosophie des 18. Jahrhunderts', *Das 18. Jahrhundert* 14:1 (1990), p.19-57.

Bollino, Fernando, *Ragione e sentimento: idee estetiche nel Settecento francese* (Bologna, 1991).

Boudri, J. Christiaan, *What was mechanical about mechanics: the concept of force between metaphysics and mechanics from Newton to Lagrange* (Dordrecht, 2002).

Brandt Bolton, Martha, 'Locke on sensory representation', in *Perception and reality: from Descartes to the present*, ed. R. Schumacher (Paderborn, 2004), p.146-67.

Bristow, William, 'Enlightenment', in *Stanford encyclopedia of philosophy* (autumn 2017 edn), ed. Edward N. Zalta, https://plato.stanford.edu/archives/fall2017/entries/enlightenment/ (last accessed 14 April 2022).

Broman, Thomas, 'Metaphysics for an enlightened public: the controversy over monads in Germany 1746-1748', *Isis* 103:1 (2012), p.1-23.

Bronisch, Johannes, *Mäzen der Aufklärung: Ernst Christoph von Manteuffel und das Netzwerk des Wolffianismus* (Berlin, 2010).

Brown, Stuart, *British philosophy and the age of Enlightenment* (London, 1996).

–, 'Leibniz and the fashion for systems and hypotheses', in

Philosophers of the Enlightenment, ed. P. Gilmour (Edinburgh, 1989), p.8-30.

Buchenau, Stefanie, 'Notions directrices et architechtonique de la métaphysique: la critique kantienne de Wolff en 1763', *Astérion* 9 (2011), doi:10.4000/asterion.2136.

–, '"L'œil ne se voit pas voir": Sulzer sur la contemplation et le sentiment de soi', *Philosophiques* 42:1 (2015), p.73-88.

–, 'Sulzers Physik der Seele zwischen Medizin und Philosophie', in *Johann Georg Sulzer: Aufklärung im Umbruch*, ed. Elisabeth Décultot, Philipp Kampa and Jana Kittelmann (Berlin, 2018), p.36-50.

Buschmann, Cornelia, 'Die philosophischen Preisfragen und Preisschriften der Berliner Akademie der Wissenschaften im 18. Jahrhundert', in *Aufklärung in Berlin*, ed. Wolfgang Förster (Berlin, 1989), p.165-227.

Buzon, Frédéric de, 'Quatre essais académiques: Johann Heinrich Lambert', *Les Cahiers philosophiques de Strasbourg* 44 (2018), p.127-48.

–, 'La question de l'espace chez Euler et sa réfraction dans la période précritique: à propos des *Réflexions sur l'espace et le temps* (1748)', *Les Cahiers philosophiques de Strasbourg* 26 (2009), p.141-56.

Calinger, Ronald S., *Leonhard Euler: mathematical genius in the Enlightenment* (Princeton, NJ, 2016).

–, 'The Newtonian–Wolffian controversy: 1740-1759', *Journal of the history of ideas* 30:3 (1969), p.319-30.

Capaldi, Nicholas, *The Enlightenment project in the analytic conversation* (Dordrecht, 1998).

Carboncini, Sonia, 'Die thomasianisch-pietistische Tradition und ihre Fortsetzung durch Christian August Crusius', in *Christian Thomasius 1655-1728: Interpretationen zu Werk und Wirkung*, ed. Werner Schneiders (Hamburg, 1989), p.287-303.

Catana, Leo, *The Historiographical concept 'system of philosophy': its origin, nature, influence and legitimacy* (Leiden, 2008).

Charles, Sébastien, 'Des excès dogmatiques à la guérison sceptique: le pyrrhonisme raisonnable de Louis de Beausobre', *Libertinage et philosophie au XVIIe siècle* 12 (2010), p.205-17.

–, 'Lumières allemandes et scepticisme: le cas de l'Académie de Berlin', in *Orthodoxie et hétérodoxie: libertinage et religion en Europe au temps des Lumières*, ed. Marie-Hélène Quéval (Saint-Etienne, 2010), p.139-49.

Charrak, André, *Contingence et nécessité des lois de la nature au XVIIIe siècle: la philosophie seconde des Lumières* (Paris, 2006).

Christensen, Thomas, 'Introduction', in *Aesthetics and the art of musical composition in the German Enlightenment*, ed. Nancy K. Baker and Thomas Christensen (Cambridge, 1995), p.3-24.

Clark, William, 'The death of metaphysics in enlightened Prussia', in *The Sciences in enlightened Europe*, ed. William Clark, Jan Golinski and Simon Schaffer (Chicago, IL, 1999), p.423-73.

Cohen, I. Bernard, *Introduction to Newton's 'Principia'* (Cambridge, 1971).

Cooper, Andrew, 'Kant and experimental philosophy', *British journal for the history of philosophy* 25:1 (2017), p.265-86.

Coopersmith, Jennifer, *The Lazy universe: an introduction to the Principle of least action* (Oxford, 2017).

Correard, Nicolas, 'Reasonable scepticism in the French Enlightenment: some connections between Jean-Baptiste Boyer d'Argens, Louis de Beausobre, and Voltaire', in *Scepticism in the eighteenth century: Enlightenment, Lumières, Aufklärung*, ed. Sébastien Charles and Plínio J. Smith (Dordrecht, 2013), p.173-88.

Couturier-Heinrich, Clémence, '"Tendance naturelle au rythme" et "observation instinctive de la mesure": l'inscription du rythme dans le domaine du spontané chez Johann Georg Sulzer et August Wilhelm Schlegel', *Revue germanique internationale* 18 (2002), p.53-70.

Crispini, Franco, 'La psicologia wolffiana in Samuel Formey e Louis de Beausobre', *Rivista di storia della filosofia* 44:3 (1989), p.477-97.

Décultot, Elisabeth, 'De la science de l'âme à la théorie des beaux-arts: l'œuvre de Johann Georg Sulzer et l'esthétique des Lumières', in *Les Lumières: un héritage et une mission – hommage à Jean Mondot*, ed. Gilbert Merlio and Nicole Pelletier (Pessac, 2012), p.177-92.

–, 'Johann Georg Sulzers System der schönen Künste', in *Johann Georg Sulzer (1720-1779): Aufklärung zwischen Christian Wolff und David Hume*, ed. Frank Grunert and Gideon Stiening (Berlin, 2011), p.211-24.

–, 'Métaphysique ou physiologie du beau? La théorie des plaisirs de Johann Georg Sulzer (1751-1752)', *Revue germanique internationale* 4 (2006), p.93-106.

–, 'Die Schattenseiten der Seele: zu Johann Georg Sulzers Theorie der dunklen Vorstellungen', in *Formen des Nichtwissens der Aufklärung*, ed. Hans Adler and Rainer Godel (Paderborn, 2010), p.263-78.

–, 'Von der Seelenkunde zur Kunsttheorie: zu Sulzers Untersuchung über den Ursprung der angenehmen und unangenehmen Empfindungen (1751/52)', *Scientia poetica* 12 (2008), p.69-88.

–, Philipp Kampa and Jana Kittelmann (ed.), *Johann Georg Sulzer: Aufklärung im Umbruch* (Berlin, 2018).

Deloche, Bernard (ed.), *L'Esthétique de Johann Georg Sulzer (1720-1779)* (Lyon, 2005).

–, 'Johann Georg Sulzer, théoricien de l'effet esthétique', in *L'Esthétique de Johann Georg Sulzer (1720-1779)*, ed. Bernard Deloche (Lyon, 2005), p.39-67.

Delon, Michel, *L'Idée d'énergie au tournant des Lumières: 1770-1820* (Paris, 1988).

di Bella, Stefano, *The Science of the individual: Leibniz's ontology of individual substance* (Dordrecht, 2005).

Dilthey, Wilhelm, 'Friedrich der Grosse und die deutsche Aufklärung', in *Studien zur Geschichte des deutschen Geistes, Gesammelte Schriften*, vol.3 (Leipzig and Berlin, 1927), p.81-157.

Duchesneau, François, 'Nicolas de Béguelin et les fondements d'une philosophie de la nature', *Philosophiques* 42:1 (2015), p.89-105.

Dugas, René, *A History of mechanics* (Mineola, NY, 2012).

Duhot, Jean-Joël, 'Sulzer et l'esthétique musicale', in *L'Esthétique de Johann Georg Sulzer (1720-1779)*, ed. Bernard Deloche (Lyon, 2005), p.103-38.

Dumouchel, Daniel, 'Entre métaphysique et esthétique: sentiment et beauté dans les mémoires de Johann Georg Sulzer', *Philosophiques* 42:1 (2015), p.51-72.

–, 'Hommes de nulle secte: éclectisme et refus des systèmes chez Jean Bernard Mérian', *Dialogue* 57:4 (2018), p.745-65.

–, 'Jean Bernard Mérian: métaphysique et empirisme à l'Académie de Berlin', in *Académies scientifiques européennes et modernité*, ed. Pierre Girard, Christian Leduc and Mitia Rioux-Beaulne (Paris, forthcoming).

–, 'Johann Georg Sulzer: la psychologie comme "physique de l'âme"', in *Les Métaphysiques des Lumières*, ed. Pierre Girard, Christian Leduc and Mitia Rioux-Beaulne (Paris, 2016), p.171-90.

–, 'Tiefen der Seele: veränderte Zustände und psychologische Paradoxe – die empirische Psychologie bei J. G. Sulzer', in *Johann Georg Sulzer: Aufklärung im Umbruch*, ed. Elisabeth Décultot, Philipp Kampa and Jana Kittelmann (Berlin, 2018), p.14-35.

Dürbeck, Gabriele, *Einbildungskraft und Aufklärung: Perspektiven der Philosophie, Anthropologie und Ästhetik um 1750* (Tübingen, 1998).

Dyck, Corey W., *Kant and rational psychology* (Oxford, 2014).

Ecole, Jean, 'Des rapports de l'expérience et de la raison dans l'analyse de l'âme ou la *Psychologia empirica* de Christian Wolff', *Giornale di metafisica* 21:4-5 (1966), p.589-617.

Erhard, Jean, *L'Idée de nature en France dans la première moitié du XVIIIe siècle* (Paris, 1963).

Euler, Werner, 'Die Idee des Schönen in Sulzers allgemeiner Theorie des Vergnügens', in *Johann Georg Sulzer (1720-1779): Aufklärung zwischen Christian Wolff und David Hume*, ed. Frank Grunert and Gideon Stiening (Berlin, 2011), p.101-33.

Florey, Ernst, and Olaf Breidbach (ed.), *Das Gehirn: Organ der Seele? Zur Ideengeschichte der Neurobiologie* (Berlin, 1993).

Fontius, Martin, '"Libertas philosophandi" und "siècle de la philosophie": zum geistesgeschichtlichen Standort Formeys', in *Franzosen in Berlin: über Religion und Aufklärung in Preußen. Studien zum Nachlass des Akademiesekretärs Samuel Formey*, ed. Martin Fontius and Jens Häseler (Basel, 2019), p.125-252.

–, and Helmut Holzhey (ed.), *Schweizer im Berlin des 18. Jahrhunderts* (Berlin, 1996).

Gaukroger, Stephen, 'The metaphysics of impenetrability: Euler's conception of force', *British journal for the history of science* 15 (1982), p.132-54.

Gay, Daniela, 'L'amitié entre Johann Georg Sulzer (1720-1779) et Charles Bonnet (1720-1793)', in *Les Ecrivains suisses alémaniques et la culture francophone au XVIII*[e] *siècle*, ed. Michèle Crogiez Labarthe, Sandrine Battistini and Karl Kürtös (Geneva, 2008), p.85-96.

Gay, Peter, *The Enlightenment: an interpretation – the rise of modern paganism* (New York, 1966).

Gesse, Sven, *Genera mixta: Studien zur Poetik der Gattungsmischung zwischen Aufklärung und Klassik-Romantik* (Würzburg, 1997).

Goldenbaum, Ursula, *Ein gefälschter Leibnizbrief? Plädoyer für seine Authentizität* (Hanover, 2016).

–, 'It is not experience that matters in the battle of rationalism and empiricism, or how theology favors empiricism over rationalism', in *Leibniz's experimental philosophy*, ed. Arnaud Pelletier, *Studia Leibnitiana* 46 (2016), p.41-68.

Gorham, Geoffrey, Benjamin Hill, Edward Slowik and C. Kenneth Waters (ed.), *The Language of nature: reassessing the mathematization of natural philosophy in the seventeenth century* (Minneapolis, MN, 2016).

Gorham, Geoffrey, and Edward Slowik, 'Locke and Newton on space and time and their sensible measures', in *Newton and empiricism*, ed. Zvi Biener and Eric Schliesser (Oxford, 2014), p.119-37.

Grau, Conrad, *Die Preussische Akademie der Wissenschaften zu Berlin* (Heidelberg, 1993).

Greenberg, John L., *The Problem of the earth's shape from Newton to Clairaut* (Cambridge, 2010).

Grunert, Frank, and Gideon Stiening (ed.), *Johann Georg Sulzer (1720-1779): Aufklärung zwischen Christian Wolff und David Hume* (Berlin, 2011).

Hagengruber, Ruth E. (ed.), *Emilie Du Châtelet between Leibniz and Newton* (Dordrecht, 2012).

Hahn, Roger, *The Anatomy of a scientific institution: the Paris Academy of sciences, 1666-1803* (Berkeley, CA, 1971).

Harnack, Adolf von, *Geschichte der Königlich Preussischen Akademie der Wissenschaften zu Berlin*, 3 vols (Berlin, 1900).

Hartz, Glenn, *Leibniz's final system: monads, matter and animals* (London, 2007).

Häseler, Jens, 'Johann Bernhard Merian – ein Schweizer an der Berliner Akademie', in *Schweizer im Berlin des 18. Jahrhunderts*, ed. Martin Fontius (Berlin, 1996), p.217-28.

–, 'Samuel Formey, pasteur huguenot entre Lumières françaises et Aufklärung', *Dix-huitième siècle* 34 (2002), p.239-47.

Hecht, Hartmut, 'Gemeinsame Denkmotive bei Leibniz und Maupertuis', in *Pierre Louis Moreau de Maupertuis: eine Bilanz nach 300 Jahren*, ed. Hartmut Hecht (Berlin, 1999), p.207-24.

Heinz, Marion, 'J. G. Sulzer und die Anfänge der Dreivermögenslehre bei Kant', in *Johann Georg Sulzer (1720-1779): Aufklärung zwischen Christian Wolff und David Hume*, ed. Frank Grunert and Gideon Stiening (Berlin, 2011), p.83-100.

Helmholtz, Hermann von, 'Rede über die Entdeckungsgeschichte des Princips der kleinsten Action', in Adolf von Harnack, *Geschichte der Königlich Preussischen Akademie der Wissenschaften zu Berlin*, 3 vols (Berlin, 1900), vol.2, p.282-96.

Hinske, Norbert, 'Die tragenden Grundideen der deutschen Aufklärung', in *Die Philosophie der deutschen Aufklärung*, ed. Raffaele Ciafardone (Stuttgart, 1990), p.407-58.

Holzhey, Helmut, 'Philosophie als Eklektik', *Studia Leibnitiana* 15:1 (1983), p.19-29.

Hornung, Erik, 'Diasporas, diversity, and economic activity: evidence from 18th-century Berlin', Centre for Competitive Advantage in the Global Economy (University of Warwick), Working Paper Series, no.390 (2018), https://warwick.ac.uk/fac/soc/economics/research/centres/cage/manage/publications/390-2018_hornung.pdf (last accessed 10 May 2022).

Howard, Stephen, 'Why did Leibniz fail to complete his dynamics?', *British journal for the history of philosophy* 25:1 (2017), p.22-40.

Hulliung, Mark, *The Autocritique of Enlightenment: Rousseau and the philosophes* (Cambridge, MA, 1994).

Hunter, Ian, *Rival Enlightenments* (Cambridge, 2001).

Israel, Jonathan, *Democratic Enlightenment: philosophy, revolution, and human rights 1750-1790* (Oxford, 2011).

Jacob, Margaret C., 'Newtonianism and the origins of the Enlightenment: a reassessment', *Eighteenth-century studies* 11:1 (1977), p.1-25.

Janiak, Andrew, *Newton as philosopher* (Cambridge, 2008).

Jourdain, Philip E. B., 'Maupertuis and the Principle of least action', *The Monist* 22 (1912), p.414-59.

Jurin, James, *The Correspondence of James Jurin (1684-1750): physician and secretary to the Royal Society*, ed. Andrea A. Rusnock (Amsterdam, 1996).

Kulstad, Mark, *Leibniz on apperception, consciousness and reflection* (Munich, 1991).

Lamblin, Annie, 'Johann Georg Sulzer', in *Aux sources de l'esthétique: les débuts de l'esthétique philosophique en Allemagne*, ed. Jean-François Goubet and Gérard Raulet (Paris, 2005), p.283-94.

Lanczos, Cornelius, *The Variational principles of mechanics* (Toronto, 1949).

Lattanzi, Lorenzo, 'Il dibattito della *Aufklärung* sull'oscurità del sentimento gradevole', in *Il secolo dei lumi e l'oscuro*, ed. Piero Giordanetti, Giambattista Gori and Maddalena Mazzocut-Mis (Milan, 2009), p.161-205.

Laursen, John C., 'Louis Isaac de Beausobre (1730-1783)', in *The Bloomsbury dictionary of eighteenth-century German philosophers*, ed. Heiner F. Klemme and Manfred Kuehn (London, 2016), p.50-52.

–, 'Tame skeptics at the Prussian Academy', *Libertinage et philosophie au XVIIe siècle* 12 (2010), p.219-28.

Leduc, Christian, 'Euler et le monadisme', *Studia Leibnitiana* 45:2 (2013), p.150-69.

–, 'La métaphysique de la nature à l'Académie de Berlin', *Philosophiques* 42:1 (2015), p.11-30.

–, 'Speculative philosophy at the Berlin Academy', in *History of universities*, vol.33.2, ed. Mordechai Feingold and Andrea Sangiacomo (Oxford, 2020), p.175-200.

–, and Daniel Dumouchel, 'Introduction', *Philosophiques* 42:1 (2015), p.7-10.

Levitin, Dmitri, 'Newton and scholastic philosophy', *British journal for the history of science* 49:1 (2016), p.53-77.

Lifschitz, Avi, *Language and the Enlightenment: the Berlin debates of the eighteenth century* (Oxford, 2012).

Lyssy, Ansgar, 'L'économie de la nature: Maupertuis et Euler sur le principe de moindre action', *Philosophiques* 42:1 (2015), p.31-51.

–, *Kausalität und Teleologie bei G. W. Leibniz* (Stuttgart, 2016).

McClellan, J. E., 'The Académie royale des sciences, 1699-1793: a statistical portrait', *Isis* 72:4 (1981), p.541-67.

McMullin, Ernan, 'The impact of Newton's *Principia* on the philosophy of science', *Philosophy of science* 38:3 (2001), p.327-38.

McRae, Robert, *Leibniz: perception, apperception, and thought* (Toronto, 1976).

Matytsin, Anton M., *The Specter of skepticism in the age of Enlightenment* (Baltimore, MD, 2016).

Metzger, Hélène, 'La théorie du feu d'après Boerhaave', *Revue philosophique de la France et de l'étranger* 109 (1930), p.253-85.

Michel, Christian, 'L'argument des origines dans les théories des arts en France à l'époque des Lumières', in *Primitivisme et mythes des origines dans la France des Lumières 1680-1820*, ed. Christian Michel and Chantal Grell (Paris, 1989), p.35-45.

Moser, Walter, 'Jean-Georges Sulzer, continuateur de la pensée sensualiste dans l'Académie de Berlin', *Modern language notes* 84 (1969), p.931-41.

Nachtomy, Ohad, 'Leibniz on nested individuals', *British journal for the history of philosophy* 15:4 (2007), p.709-28.

Nannini, Alessandro, 'At the bottom of the soul: the psychologization of the *fundus animae* from Leibniz to Sulzer', *Journal of the history of ideas* 82:1 (2021), p.51-72.

–, 'Umanizzare l'uomo: le implicazioni antropologiche dell'educazione estetica nell'epoca della *Popularphilosophie*', *Cultura tedesca* 50 (2016), p.101-17.

Paccioni, Jean-Paul (ed.), 'Présentation', *Astérion* 9 (2011), special issue: *La Preisfrage de 1763 de l'Académie de Berlin: certitude mathématique, certitude métaphysique, certitude morale*, ed. Jean-Paul Paccioni, https://doi.org/10.4000/asterion.2090.

Pallenberg, Olivia, 'Immaginazione e follia: una discussione all'Accademia di Berlino', in *Dal Cartesianismo all'Illuminismo radicale*, ed. Carlo Borghero and Claudio Buccolini (Florence, 2010), p.143-58.

Palme, Anton, *Sulzers Psychologie und die Anfänge der Dreivermögenslehre* (Berlin, 1905).

Panza, Marco, 'De la nature épargnante aux forces généreuses: le principe de moindre action entre mathématiques et métaphysique – Maupertuis et Euler, 1740-1751', *Revue d'histoire des sciences* 48:4 (1995), p.435-520.

Petzoldt, Joseph, 'Maxima, Minima und Oekonomie', *Vierteljahresschrift für wissenschaftliche Philosophie* 14 (1890), p.206-39, 354-66 and 417-42.

Pirro, Maurizio, 'Sulzers Physik der Seele und die Dramentheorie Schillers', *Zeitschrift für Germanistik* 16 (2006), p.314-23.

Plett, Heinrich Franz, *Rhetorik der Affekte: Englische Wirkungsästhetik im Zeitalter der Renaissance* (Tübingen, 1975).

Pott, Sandra, Martin Mulsow and Lutz Danneberg (ed.), *The Berlin Refuge, 1680-1780: learning and science in European context* (Leiden, 2003).

Proß, Wolfgang, '"Meine einzige Absicht ist, etwas mehr Licht über die Physik der Seele zu verbreiten": Johann Georg Sulzer (1720-1779)', in *Helvetien und Deutschland: Kulturelle Beziehungen zwischen der Schweiz und Deutschland in der Zeit von 1770-1830*, ed. Hellmut Thomke, Martin Bircher and Wolfgang Proß (Amsterdam, 1994), p.134-48.

Prunea-Bretonnet, Tinca, 'La méthode philosophique en question: l'Académie de Berlin et le concours pour l'année 1763', *Philosophiques* 42:1 (2015), p.107-30.

–, 'On the mitigated phenomenalism of J.-B. Merian', in *The Experiential turn in eighteenth-century German philosophy*, ed.

Karin de Boer and Tinca Prunea-Bretonnet (New York, 2021), p.203-27.

–, 'Prouver l'existence de Dieu à partir de l'observation de la nature: Formey (et Wolff) sur l'argument des causes finales', in *Berliner Aufklärung: Kulturwissenschaftliche Studien*, ed. Ursula Goldenbaum and Alexander Košenina, vol.7 (Hanover, 2020), p.165-83.

Pugh, David, 'Aesthetic humanism and its foes: the perspective from Halle', in *Who is this Schiller now? Essays on his reception and significance*, ed. Jeffrey L. High, Nicholas Martin and Norbert Oellers (Rochester, NY, 2011), p.116-32.

Pulte, Helmut, *Axiomatik und Empirie: eine wissenschaftstheoriegeschichtliche Untersuchung zur Mathematischen Naturphilosophie von Newton bis Neumann* (Darmstadt, 2005).

–, *Das Prinzip der kleinsten Wirkung und die Kraftkonzeptionen der rationalen Mechanik* (Stuttgart, 1989).

Rey, Anne-Lise, 'Le leibnizo-newtonianisme: la construction d'une philosophie naturelle complexe dans la première moitié du 18e siècle: la méthode d'Emilie Du Châtelet entre hypothèses et expériences', *Dix-huitième siècle* 45:1 (2013), p.115-29.

Riedel, Wolfgang, 'Erkennen und Empfinden: Anthropologische Achsendrehung und Wende zur Ästhetik bei Johann Georg Sulzer', in *Der ganze Mensch: Anthropologie und Literatur im 18. Jahrhundert*, ed. Hans-Jürgen Schings (Stuttgart, 1994), p.410-39.

Riley, Matthew, *Musical listening in the German Enlightenment: attention, wonder and astonishment* (Aldershot, 2004).

Roger, Jacques, *Les Sciences de la vie dans la pensée française au XVIIIe siècle*, 3rd edn (Paris, 1993).

Rojo, Alberto, and Anthony Bloch, *The Principle of least action, history and physics* (Cambridge, 2018).

Schliesser, Eric, 'Newton and Newtonianism', in *The Routledge companion to eighteenth-century philosophy*, ed. Aaron Garrett (New York, 2014), p.62-90.

Schneider, Ulrich J., 'The problem of eclecticism in the history of philosophy', *Intellectual history review* 26:1 (2016), p.117-29.

Schneiders, Werner (ed.), *Lexikon der Aufklärung: Deutschland und Europa* (Munich, 2001).

–, 'Vernünftiger Zweifel und wahre Eklektik', *Studia Leibnitiana* 17:2 (1985), p.143-61.

–, *Die wahre Aufklärung: zum Selbstverständnis der deutschen Aufklärung* (Freiburg and Munich, 1974).

–, 'Zwischen Welt und Weisheit: zur Verweltlichung der Philosophie in der frühen Moderne', in *Philosophie der Aufklärung – Aufklärung der Philosophie: Gesammelte Schriften* (Berlin, 2005), p.343-64.

Scholl, Ann, *Descartes's dreams: imagination in the Meditations* (New York, 2005).

Shank, J. B., *The Newton wars and the beginning of the French Enlightenment* (Chicago, IL, 2008).

Shapiro, Alan E., 'Newton's experimental philosophy', *Early science and medicine* 9:3 (2004), p.185-217.

Sheynin, Oscar B., 'J. H. Lambert's work on probability', *Archive for history of exact sciences* 7:3 (1971), p.244-56.

Stiening, Gideon, 'Zur physischen Anthropologie einer Unsterblichkeit der Seele', in *Johann Georg Sulzer (1720-1779): Aufklärung zwischen Christian Wolff und David Hume*, ed. Frank Grunert and Gideon Stiening (Berlin, 2011), p.57-81.

Stöckmann, Ernst, *Anthropologische Ästhetik: Philosophie, Psychologie und ästhetische Theorie der Emotionen im Diskurs der Aufklärung* (Tübingen, 2009).

–, 'Von der sinnlichen Erkenntnis zur Psychologie der Emotionen', in *Physis und Norm: neue Perspektiven der Anthropologie im 18. Jahrhundert*, ed. Manfred Beetz, Jörn Garber and Heinz Thoma (Göttingen, 2007), p.69-106.

Suchting, Wallis A., 'Euler's reflections on space and time', *Scientia: an international review of scientific synthesis* 104 (1969), p.270-78.

Suisky, Dieter, *Euler as physicist* (Dordrecht, 2009).

–, 'Euler's mechanics as a foundation of quantum mechanics', in *Leonhard Euler: life, work and legacy*, ed. Robert E. Bradley and Charles Edward Sandifer (Amsterdam, 2007), p.503-25.

Terrall, Mary, 'The culture of science in Frederick the Great's Berlin', *History of science* 28 (1990), p.333-64.

–, *The Man who flattened the earth: Maupertuis and the sciences in the Enlightenment* (Chicago, IL, 2002).

Terras, Rita, 'The development of the doctrine of the three faculties in German aesthetics', *Monatshefte* 70:1 (1978), p.5-14.

Thiel, Udo, 'Between Wolff and Kant: Merian's theory of apperception', *Journal of the history of philosophy* 34 (1996), p.213-32.

–, *The Early modern subject: self-consciousness and personal identity from Descartes to Hume* (Oxford, 2014).

–, 'Sulzer über Bewusstsein im Kontext', in *Johann Georg Sulzer (1720-1779): Aufklärung zwischen Christian Wolff und David Hume*, ed. Frank Grunert and Gideon Stiening (Berlin, 2011), p.21-36.

Tonelli, Giorgio, 'La disputa sul metodo matematico nella filosofia della prima meta del settecento e la genesi dello scritto Kantiano sull'evidenza', in *Da Leibniz a Kant: saggi sul pensiero del settecento*, ed. Claudio Cesa (Naples, 1987), p.81-107.

–, *La Pensée philosophique de Maupertuis* (Hildesheim, 1987).

Torra-Mattenklott, Caroline, *Metaphorologie der Rührung: Ästhetische Theorie und Mechanik im 18. Jahrhundert* (Munich, 2002).

Van der Zande, Johan, 'Johann Georg Sulzer – Spaziergänge im Berliner Tuskulum', in *Berliner Aufklärung: Kulturwissenschaftliche Studien*, ed. Ursula Goldenbaum and Alexander Košenina, vol.1 (Hanover, 1999), p.41-68.

–, 'Orpheus in Berlin: a reappraisal of Johann Georg Sulzer's theory of polite arts', *Central European history* 28 (1995), p.175-208.

–, 'Thespis for and against: Sulzer and Rousseau on theater and politics', in *Reconceptualizing nature, sciences and aesthetics*, ed. Patrick Coleman (Geneva, 1998), p.231-46.

Vesper, Achim, 'Le plaisir du beau chez Leibniz, Wolff, Sulzer, Mendelssohn et Kant', *Revue germanique internationale* 4 (2006), p.23-36.

–, 'Sulzer über die schönen Künste und das Gute', in *Johann Georg Sulzer (1720-1779): Aufklärung zwischen Christian Wolff und David Hume*, ed. Frank Grunert and Gideon Stiening (Berlin, 2011), p.169-89.

Vittadello, Anna M., 'Expérience et raison dans la psychologie de Christian Wolff', *Revue philosophique de Louvain* 71 (1973), p.488-511.

Westfall, Richard, *The Construction of modern science* (Cambridge, 1971).

Woolhouse, Roger, 'Locke's idea of spatial extension', *Journal of the history of philosophy* 8:3 (1970), p.313-17.

Wunderlich, Falk, 'Johann Georg Sulzers Widerlegung des Materialismus und die Materietheorien der Zeit', in *Johann Georg Sulzer (1720-1779): Aufklärung zwischen Christian Wolff und David Hume*, ed. Frank Grunert and Gideon Stiening (Berlin, 2011), p.37-55.

Wundt, Max, *Die deutsche Schulphilosophie im Zeitalter der Aufklärung* (1945; Hildesheim, 1992).

Yourgrau, Wolfgang, and Stanley Mandelstam, *Variational principles in dynamics and quantum theory*, 2nd edn (New York, 1960).

Zelle, Carsten, 'Encyclopédisme et esthétique: la dimension européenne de Sulzer', in *Ferments d'ailleurs: transferts culturels entre Lumières et romantismes*, ed. Denis Bonnecase and François Genton (Grenoble, 2010), p.187-212.

Index